AP

Advanced Placement

Biology

Tamar Aprahamian, Ph.D
Claudine Land, PhD

XAMonline, Inc.

XAMonline, Inc.
21 Orient Avenue
Melrose, MA 02176
Toll Free: 1-800-509-4128
Email: info@xamonline.com
Web: www.xamonline.com
Fax: 1-617-583-5552

Library of Congress Cataloging-in-Publication Data
Aprahamian, Tamar

AP Biology/ Tamar Aprahamian
ISBN: 978-1-60787-637-3

1. AP   2. Study Guides   3. Science   4. Biology

Disclaimer:
The opinions expressed in this publication are the sole works of XAMonline and were created independently from The College Board, or other testing affiliates. Between the time of publication and printing, specific test standards as well as testing formats and website information may change that are not included in part or in whole within this product. XAMonline develops sample test questions, and they reflect similar content as on real tests; however, they are not former tests. XAMonline assembles content that aligns with test standards but makes no claims nor guarantees candidates a passing score.

Cover photos provided by © istock.com/Alberto Masnovo/54300968, istock.com/Andrey Prokhorov/21506562, istock.com/photka/9323746, CanStockPhoto/Eraxion/2309923, CanStockPhoto/SergeyNivens/10622103

Printed in the United States of America

AP Biology
ISBN: 978-1-60787-637-3

Table of Contents

Dr. Aprahamian received her Ph.D. in Cell, Molecular, and Developmental Biology from Tufts University in Boston, Massachusetts. She currently runs a research laboratory focused on understanding the integrated molecular pathways of cardiovascular and metabolic complications in autoimmunity and obesity. She has been awarded competitive NIH and foundation grants, as well as industry-sponsored research awards for preclinical studies. She has published more than 30 papers in peer-reviewed scientific journals, and serves as a reviewer for grant study sections and academic journals. Dr. Aprahamian is also the founder of JetPub Scientific Editing Services, which provides content writing and English language editing for scientific and medical documents.

Dr. Land holds a Ph.D. in molecular and cellular biology from Tulane University, with postdoctoral research experience in molecular entomology, noncoding RNA, and computational genomics. She has published several peer reviewed journal articles and academic book chapters, as well as numerous popular science articles.

SECTION I:
Introduction

Chapter I:
What To Expect In This Book

Using this Study Guide

This study guide is organized according to the Big Ideas set forth by the College Board. The content assumes that you know the basics from an introductory biology course.

You are already familiar with the major biology concepts that you should know, and this book specifically relates them to the College Board concepts that they want you to understand. Within the margins, look for general statements about biology. These are direct quotes from the College Board's AP Biology Course and Exam Description (Revised Edition, effective Fall 2015, The College Board, New York, NY). In other words, these are the exact concepts that the makers of the AP Biology exam want you to know and understand for the test. Make sure you read these carefully, and if there are any topics that seem unfamiliar, note these as areas for further study. Furthermore, at the beginning of each chapter, a few bullets will outline some key points that the chapter will emphasize.

Each chapter is broken down by main topic from the Big Ideas set forth by the College Board and includes sample multiple choice questions for that topic. While these are not necessarily in the same format as those found on the AP Biology exam, they will help to assess your knowledge of the different concepts. You will also find two, full-length sample tests at the end of this book. These are designed to give you hands-on experience that simulates the actual exam you will be taking. Answers are given for each of the sample questions as well as detailed rationale, points to remember, and pitfalls to watch out for. Use this information to help guide your learning.

About the Advanced Placement Biology Program

The Advanced Placement® program is designed to offer students college credit while still in high school. The more than 30 AP courses culminate in an intensive final exam given every year in May.

AP Biology covers the life science material that would be offered in a full year freshman-level biology course. The curriculum contains both the theoretical and conceptual side of biology as well as a strong lab component. It is through both of these that students develop a deep understanding of biological processes and how to think like scientists.

Successful completion of the course and a passing score on the exam not only provides students with a deep sense of accomplishment, but also gives them a jumpstart on their college careers. AP credit is almost universally accepted by post-secondary schools, however each school has different guidelines as to what scores they will accept.

About the New Exam

When the Advanced Placement Biology course was originally designed, it covered a vast array of biological topics ranging from lipids to predators, from mitochondria to the desert. Students were basically required to know everything that was presented in a freshman-level text. The argument was that the AP Biology curriculum was a "mile wide and an inch deep." Students needed to know a lot of material, but in very little depth. Students not only found this to be overwhelming, but teachers also had difficulty fitting in an entire textbook's worth of material between September and May. There was just far too much to know, and that's why many teachers implemented summer assignments to give students a head start.

In 2012, the AP Biology curriculum was redesigned so that fewer concepts were presented, and those that were still included were examined in much greater depth. There is still a lot of content to cover (hey, biology is a complicated subject!), but the amount of detail students need to know has been reduced. Additionally, the investigative portion of the course has been ramped up. Students are asked to spend a large portion of their course doing lab work, creating their own hypotheses and designing experiments.

The New Curriculum

The major revision that took place in the AP Biology curriculum was organizing all of the content around four underlying principles called "Big Ideas." As described below, these Big Ideas encompass scientific principles, theories, and processes that cut across traditional boundaries and provide a broad way of thinking about all of the biology underlying the observations you make about the physical world.

1. *The Process of Evolution* – this is the cornerstone of everything that happens in the living world. It includes analyses of evolution's origins (Darwin, Lamarck) and discusses current understandings of the process (cladograms, phylogenies, evolutionary development). All of the other Big Ideas can be connected back to this one.

2. *Living Things Use Energy and Molecular Components for Growth, Metabolism, and Reproduction* – Organismal and cellular growth and reproduction is founded on the organization of living systems that require free energy and matter to maintain dynamic homeostasis.

3. *Information is Stored, Retrieved, Transmitted, and Responded To by Living Things* – DNA is the prime component of living things that contains all of the information they need to survive and reproduce. For living systems to store, retrieve, transmit and respond to information essential to life processes, we must understand the heritable continuity that is fundamental to all life.

4. *Interactions Among Living Things are Complex* – Living things interact with each other. This may be two paramecia competing for a single food source or two male lions fighting over the right to mate with a female. At the cellular level, cells often work together to process stimuli and maintain homeostasis. Cellular communication is very important for maintaining a constant internal system.

The Make Up of the Exam

The AP Biology exam is administered every May. It lasts for about three hours and is made of the following parts:

Part I – Objective Questions – 90 minutes

    A. Multiple Choice – 63 questions

    B. Grid-In – 6 questions

Part II – Free Response – 80 minutes + 10 minutes preparation period

    A. Long Response – 2 questions

    B. Short Response – 6 questions

As mentioned above, the new curriculum involves more basic laboratory skills and scientific thinking as compared to the old one. An example of an old-style question would be:

Which organelle is responsible for the energy cells use to carry out homeostasis?

(A) mitochondria

(B) lysosomes

(C) nucleus

(D) endoplasmic reticulum

Answer: A

As you can see, this is a very factual item. You either know the answer or you do not.

The new-style questions are a bit more involved. Here is an example for a new style question:

Adiponectin is a cytokine produced by fat cells that exerts anti-inflammatory effects in part by decreasing the amount of the pro-inflammatory cytokine TNF-alpha. The drug pioglitazone increases the circulating level of adiponectin and therefore is thought to be beneficial for decreasing the severity of the autoimmune disease lupus. Which graph below shows the relationship between adiponectin and disease activity after administration of the pioglitazone?

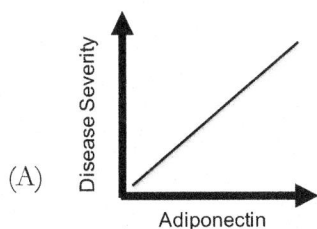

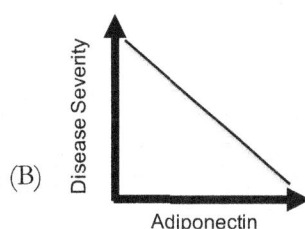

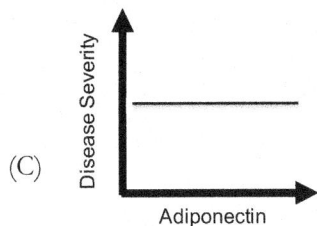

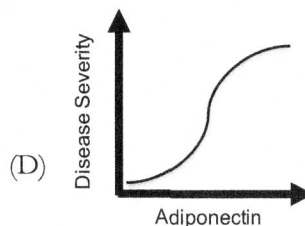

Answer: B

Compared to the previous style, you can see that this question integrates more knowledge into one question, and therefore requires more thought!

A note here about the math on the new exam. The bad news is that, yes, there is math and you will most likely have to do some calculations. The good news is that they will give you the needed formulas and you can use a calculator.

Free response questions are just that, free. There are no right or wrong answers, just good and bad ones. These questions can range in topic from cellular metabolism to ecology. There is also almost always at least one question pertaining to one of the laboratory exercises you would have performed. There will be more on how to prepare and answer the free response questions later on.

How the Exam is Scored

The multiple choice part of the test is scored by machine and the free response portion is scored by hand (every summer hundreds of professors, content specialists, and AP Biology teachers meet to grade the 300,000+ exams that are taken). Once both scores have been tallied, they are combined and then scaled. This raw score is then changed into a composite score ranging from 1 – 5.

The College Board proposes the following qualifications for each of the potential score:

Exam Grade	Recommendation
5	Extremely Well Qualified
4	Well Qualified
3	Qualified
2	Possibly Qualified
1	No recommendation

The minimum score required for college credit to be granted is a 3. As mentioned above, many schools require scores of 4 or 5 in order to grant credit.

For comparison, the College Board makes the equivalents of the AP Exam scores at follows:

AP Exam Grade	Letter Grade Equivalent
5	A
4	A-, B+, B
3	B-, C+, C
2	None
1	None

For reference, the 2015 administration of the AP Biology Exam had this distribution:

Exam Score	Percentage of Students
5	6.2
4	22
3	35.9
2	27.6
1	8.3

Based upon these scores, the distribution of students' grades would form a typical bell curve.

Hints for Taking the Test

Part I – Objective Questions (90 minutes)

Multiple choice questions can be tricky. Many times it is possible to eliminate one or two of the answers right away, but then get stuck with the remaining ones. On the AP Biology exam, there is no penalty for incorrect answers, so be sure to record an answer for every question, even if it is a complete guess.

It is also very important to know what the question is asking of you. The College Board is notorious for saying things like, "All of the following are examples, EXCEPT…" or, "Which of these is NOT…" These words can change the entire meaning of the sentence. Be on alert for qualifiers like this.

You will be using a number 2 pencil to bubble in your answers on an answer sheet. At this stage in your academic career, you have taken enough tests of this type that hopefully you know how to properly fill in the circles. If you need to erase an answer, be sure to do it completely.

Be sure of your timing. Sure, 69 questions (63 multiple choice and 6 grid-ins) in 90 minutes sounds easy enough. But remember, some of these questions are going to take longer to answer. The grid-ins often require calculations, which will eat away at your time.

Part II – Free Response (80 minutes + 10 minute preparation)

The free response questions are usually the items that give students the most difficulty. This is not because they do not know the answers. The problem usually arises from not organizing one's thoughts sufficiently and not putting them down on paper fast enough. There are two long response questions and six shorter ones. Try to devote 20 minutes to each of the long ones and only about 6 minutes for each of the short ones.

Another change that has occurred in the past years is the addition of a 10-minute preparation period. It is during this time that you can read the exam questions and start putting your thoughts in order. You can even make an outline if you wish. Be sure to write down any key terms that will be important.

The free response questions are graded on a point scale, with different key topics being worth different amounts. The maximum a question can be worth is 10 points (for example, of the question has two parts, each part is worth five points. If the question has three parts, then each part would be worth three points and the extra point would be given if all three parts are answered completely). In order to get full credit for each question, your answer must provide enough details that the reader believes you have a complete understanding of the topic. The free response questions are often several paragraphs long and can be organized into main talking points of the question that concludes with summaries for each point.

Like the multiple choice, you do not lose credit for presenting incorrect information. However, you do lose time. Other things the readers do not care about include: spelling, grammar, and penmanship. Obviously, if a reader in unable to decipher your writing, they cannot grade it, but they do their best to interpret a student's "chicken scratch."

Be certain your writing is in essay form (tell a story). Do NOT just list important concepts in an outline.

Also like the multiple choice questions, the free response questions have key terms to which you should pay particular attention. These terms include, "Compare," "Contrast,"

"Describe," and their favorite, "Explain." Pay particular attention to these terms and be sure to do exactly what they ask.

Be coherent in your writing. You do not want to say one thing in the first paragraph and then say the complete opposite in the second paragraph. If you do this, you will not get any credit, even if one of the statements is correct. This is because the reader does not know if you knew which one was correct or just took a guess and got lucky.

It is acceptable to include diagrams or graphs in your answers. If you do, be sure to properly number the visual so the reader knows to which question or part of the question it belongs. Additionally, make certain to label everything, such as axes on the graphs and units!! The last thing you want is to lose points because your beautifully constructed graph or data table was missing a title.

Finally, the biggest piece of advice for answering the free response questions is to answer the question and then move on. Do not spend time going back over it to edit it and turn it into a major piece of literature (while certainly reread it to make sure it makes sense). You do not have time for this. Write what the question asks you to write and then move on. This will be the fastest 90 minutes you have ever experienced and you have a lot, a LOT of information to get through, so writing brief, but detailed, essays is essential.

Great Study and Testing Tips!

This study guide is focusing on *what* to study in order to prepare for the subject assessments but it is equally important *how* you study.

You can increase your chances of truly mastering the information by taking some simple, but effective steps.

1. **Certain foods aid the learning process.** Foods such as milk, nuts, seeds, rice, and oats help your study efforts by releasing natural memory enhancers called CCKs (*cholecystokinin*) composed of *tryptophan, choline,* and *phenylalanine.* All of these chemicals enhance the neurotransmitters associated with memory. Before studying, try a light, protein-rich meal of eggs, turkey, and fish. All of these foods release memory enhancing chemicals. The better the connections, the more you comprehend.

   Likewise, before you take a test, stick to a light snack of energy boosting and relaxing foods. A glass of milk, a piece of fruit, or some peanuts all release various memory-boosting chemicals and help you to relax and focus on the subject at hand.

2. **Learn to take great notes.** A by-product of our modern culture is that we have grown accustomed to getting our information in short doses (i.e., TV news sound bites or USA Today style newspaper articles). Consequently, we've subconsciously trained ourselves to assimilate information better in neat little packages. If your notes are scrawled all over the paper, it fragments the flow of information. Strive for clarity. Newspapers use a standard format to achieve clarity. Your notes can be much clearer through use of proper formatting. A very effective format is called the *"Cornell Method."*

   Take a sheet of loose-leaf lined notebook paper and draw a line all the way down the paper about 1–2" from the left-hand edge.

Draw another line across the width of the paper about 1–2" from the bottom. Repeat this process on the reverse side of the page.

Look at the highly effective result. You have ample room for notes, a left hand margin for special emphasis items or inserting supplementary data from the textbook, a large area at the bottom for a brief summary, and a little rectangular space for just about anything you want.

3. **Get the concept, then the details.** Too often we focus on the details and don't gather an understanding of the concept. However, if you simply memorize only dates, places, or names, you may well miss the whole point of the subject.

A key way to understand concepts is to express them in your own words. If you are working from a textbook, automatically summarize each paragraph in your mind. If you are outlining text, don't simply copy the author's words.

Rephrase them in your own words. You remember your own thoughts and words much better than someone else's, and subconsciously tend to associate the important details to the core concepts.

4. **Ask Why?** Pull apart written material paragraph by paragraph and don't forget the captions under the illustrations.

Example: If the heading is "Stream Erosion," flip it around to read "Why do streams erode?" Then answer the questions.

If you train your mind to think in a series of questions and answers, not only will you learn more, but it also helps to lessen the test anxiety because you are used to answering questions.

5. **Read for reinforcement and future needs.** Even if you only have 10 minutes, put your notes or a book in your hand. Your mind is similar to a computer; you have to input data in order to have it processed. *By reading, you are creating the neural connections for future retrieval.* The more times you read something, the more you reinforce the learning of ideas.

Even if you don't fully understand something on the first pass, *your mind stores much of the material for later recall.*

6. **Know your strengths**. Our bodies respond to an inner clock called the biorhythm. Burning the midnight oil works well for some people, but not for everyone.

If possible, set aside a particular place to study that is free of distractions. Shut off the television, cell phone, and pager and exile your friends and family during your study period.

If you really are bothered by silence, try some background music. Light classical music at a low volume has been shown to aid in concentration over other types. Music that evokes pleasant emotions without lyrics is highly suggested. Try just about anything by Mozart. It will relax you.

7. **Use arrows not highlighters.** At best, it's difficult to read a page full of yellow, pink, blue, and green streaks. Try staring at a neon sign for a while and you'll soon see that the horde of colors obscures the message.

A quick note, a brief dash of color, an underline, and an arrow pointing to a particular passage is much clearer than a horde of highlighted words.

8. **Budget your study time.** Although you shouldn't ignore any of the material, *allocate your available study time in the same ratio according to how the topics may appear on the test.*

Test Taking Tips:

1. **Be smart, play dumb. Don't read into the question.** Don't make an assumption that the test writer is looking for something other than what is asked. Stick to the question as written and don't read into it.

2. **Read the question and all the choices *twice* before answering the question.** You may miss something by not carefully reading, and then re-reading, both the question and the answers.

   If you really don't have a clue as to the right answer, leave it blank on the first time through. Go on to the other questions, as they may provide a clue as to how to answer the skipped questions.

   Go through all the ones you know and then go back and review the ones you have skipped. If later on, you still can't answer the skipped ones . . . **Guess.** *You may be able to narrow down the choices and improve your odds. There is no penalty for a wrong answer.* Only one thing is certain; if you don't put anything down, you will get it wrong! Do not leave any questions blank!

3. **Turn the question into a statement.** Look at the way the questions are worded. The syntax of the question usually provides a clue. Does it seem more familiar as a statement rather than as a question? Does it sound strange?

   By turning a question into a statement, you may be able to spot if an answer sounds right, and it may also trigger memories of material you have read.

4. **Look for hidden clues.** It's actually very difficult to compose multiple-choice questions without giving away part of the answer in the options presented.

   In most multiple-choice questions, you can often readily eliminate one or two of the potential answers. This leaves you with only two real possibilities and automatically your odds go to fifty-fifty for very little work.

5. **Trust your instincts.** For every fact that you have read, you subconsciously retain something of that knowledge. On questions that you aren't really certain about, go with your basic instincts. **Your first impression on how to answer a question is usually correct.**

6. **Mark your answers directly on the test booklet.** Don't bother trying to fill in the optical scan sheet on the first pass through the test.

   Just be very careful not to mis-mark your answers when you eventually transcribe them to the scan sheet.

7. **Remember you are NOT permitted to use a calculator on the multiple choice section.** Therefore, most of the problems will involve calculations that are easy to do "by hand". If you find your answer is requiring you to do complex math that you cannot do easily without a calculator this is a clue that perhaps you are handling the calculations wrong. Go back and quickly review the problem and see if there are any assumptions you can make to simplify the calculation.

8. **<u>Watch the clock!</u>** You have a set amount of time to answer the questions. Don't get bogged down trying to answer a single question at the expense of 10 questions you can more readily answer.

Chapter II:
Science Practices

A. The Scientific Method

Science progresses through the use of the scientific method. This section consists of a summary of the steps that are used in pursuing the scientific research. This method is used in all disciplines of science: biology, chemistry, physics, geology, and other hard sciences.

To understand the biological world, we use two mechanisms: our sensory perception and our ability to reason. We can identify and count the types and number of plants in a forest or identify birds by listening to their calls. With the information we gather from our senses, we can use logic and reason to infer that we would not find pink flamingos in North American forests. Logic and reason allows us to make predictions about the natural world and make predictions about the future. The ability of scientists to make predictions accurately hinges on a collection of steps called the Scientific Method, which is a universal way that scientists test their ideas. The scientific method has 8 steps:

1. State the problem
2. Collect background information
3. Establish a hypothesis
4. Perform the experiment
5. Analyze the data
6. Repeat the experiment
7. Draw conclusions
8. Report the results

B. Processes involved in scientific inquiry

Let us now go step by step through the scientific method to get a better understanding of how each step informs the next.

Step 1. State the Problem

Here you want to identify some phenomenon that occurs in nature and ask a question about it. You might want to know how elephants select their mates, or how a certain new investigational medication will affect circulating cholesterol levels. Most likely, this problem will arise from some form of observation that has been made.

Step 2. Collect Background Information

Once you have identified the problem, you will want to find out if anyone has already asked that question and performed experiments to answer it. Now is the time to research what has already been reported. You might find that other researchers have already studied your problem and answered the question. If that is the case, you could accept what they have reported, or you could extend their research findings.

For example, you want to know how long it takes chicken eggs to hatch at 37°. If your background research results in several studies concluding that the eggs hatch in 21 days, you might decide to determine if a change in temperature would affect the outcome.

Step 3. Establish a hypothesis

Using our ability to reason, we can use observations to predict future events or patterns in biology. Prediction uses inductive reasoning—deriving a generalization from specific details—to form a general principle or explanation, called a **hypothesis**.

Hypotheses have the following characteristics:

- A hypothesis states a general principle that holds up to testing across space and time.
- A hypothesis is a tentative idea.
- It should not contradict available observations.
- It should be kept as simple as possible, with the most logical explanation possible.
- It is testable and potentially falsifiable.
- Note: A hypothesis with cause-and-effect relationships has a hypothesis that asserts no effect. This is a "null hypothesis."

Example

Good: "All bacteria have flagella" would be a good hypothesis. We can test it through observation. Looking at bacteria, we would find that there are some that do not have flagella, so we would have shown our hypothesis to be false.

Bad: "Some bacteria have flagella" would be a useless hypothesis as there is no observation that would not fit this hypothesis.

Null: "The drug Atorvastatin does not lower circulating cholesterol levels."

Step 4. Perform the experiment

To test the hypothesis we must design and perform an experiment based on our prediction so that we can use our sensory perception to collect information. A good experiment takes into account the conditions that are subject to change, called variables. To compare an effect in an **experimental** group subjected to the variable we want to test, we must include a **control** group in the experiment that is not subjected to the same variable.

Example

- Hypothesis: The drug Atorvastatin does not lower circulating cholesterol levels.

- Experiment: 1400 patients between the ages of 65–75 with high cholesterol are randomly assigned to two groups of 700 patients. All patients will have their blood taken before medication is administered, and cholesterol levels will be measured. One group will take Atorvastatin once per day (experimental group), and the other group will take a placebo once per day (control group). Both groups will take a daily pill for four months, at the end of which their circulating cholesterol levels will be measured to determine any changes.

Step 5. Analyze the data

Results are collected and then subjected to statistical analysis to determine if any differences that might have been observed are significant or not.

Example

- In the Atorvastatin experiment we observed the following results: 587 patients had reduced cholesterol levels after taking Atorvastatin, while 96 patients in the control group had reported improvement.
- The data appears to show a difference between the group of patients that took Atorvastatin and those that did not.
- After testing the results with statistical analysis we determine that there was a significant effect of Atorvastatin in lowering high cholesterol levels.

Step 6. Repeat the experiment

You will recall that from earlier that all hypotheses need to be testable. In the same way, all the experiments you perform must be repeatable. This means that any researcher should be able to replicate the experiment exactly the same way that you originally performed it using your experimental design. This is used to validate and give credibility to your results.

In addition, it is important to increase the internal validity of the results. If the experiment is only performed once, then it is highly likely that the findings could just be due to chance. However, if the experiment was repeated several times, and the results are similar almost every time, then there is good reason to believe that your results are true.

Step 7. Draw a conclusion

After the results of the experiment are analyzed, a conclusion can be drawn based on the outcome. We would compare the results and determine if they agree or disagree with what we predicted and thus would allow us to accept or reject our hypothesis. It is important to understand that the experiment does not prove a hypothesis to be correct, but it can disprove a hypothesis. Instead we fail to disprove a hypothesis.

If the experiment disproves the hypothesis, we would change our hypothesis based on our new observations and test it again. If we fail to disprove the hypothesis, repeating the experiment, or additional experiments would be designed to confirm it.

Example

- In the Atorvastatin example, we can reject the null hypothesis because the patients treated with the drug did have reduced cholesterol levels.
- Revised hypothesis: The administration of Atorvastatin reduces high cholesterol compared to the administration of a placebo.
- With present information we accept the revised hypothesis to be true.
- This is not a proven hypothesis, because there could always be unaccounted variables that were the cause of the decreased cholesterol.

Step 8. Report your results

Scientists publish their findings in scientific journals and books. They also present their work publically and in meetings with other scientists. By disseminating their results, they are participating in the most essential part of the scientific method. It allows other scientists to verify and test your hypotheses, potentially develop new tests, and even apply the knowledge you have gained to solve other problems. The new results that scientists observe are put into the context of previous knowledge: what was known before on this particular subject and how the new findings compare to the previous results.

The summary of the steps are outlined in the following figure:

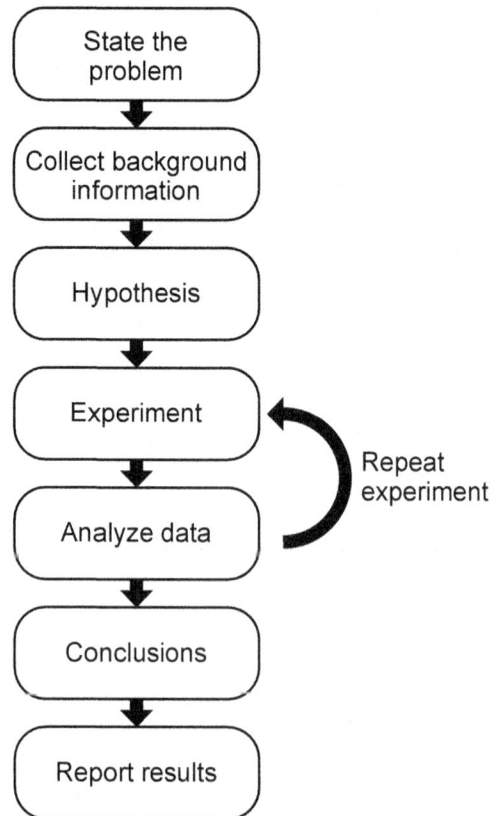

State the problem
↓
Collect background information
↓
Hypothesis
↓
Experiment
↓
Analyze data
→ Repeat experiment
↓
Conclusions
↓
Report results

Scientific inquiry is a valuable strategy for answering questions based on observations. This flow chart represents the important steps taken by scientists to answer questions.

C. Process skills

Now that you have gone through the steps of scientific inquiry, let's practice. Suppose we pose the question "Is there a relationship between where a student sits in class and their grade on the final exam?" To pose this question as a hypothesis: "Students who sit in the front three rows in class earn higher grades on average than the students sitting in the back three rows".

Three important things to remember about hypotheses are:

- A hypothesis is consistent with existing observations and known information regarding the question.
- A hypothesis must be presented as a statement of the predicted outcome, not as a question.
- A hypothesis must be specific and testable.

Try for yourself.

Write three questions based on observations you can make in the room you are sitting in.

1.
2.
3.

Now that you have three questions write them in the form of a hypothesis.

1.
2.
3.

D. Experimental design.

In designing a controlled experiment the scientist must have at least two setups: the experimental setup(s) that receives the test variable; and a control setup that does not receive the test variable. The test variable in the experimental setup is known as the **independent** variable. The two setups must be identical except for the independent variable so that the investigator is able to attribute changes between the two groups, the dependent variable, to the test treatment. All of the factors that are kept the same in the experimental and control setups are called **standardized** variables.

One important component of good experimental design is that sample size and replication of experiments is important to ensure higher levels of confidence in the results. Small sample sizes are less reliable then larger sample sizes; a single experimental result is less reliable than the same result observed several times in replicated experiments.

Now that you have three hypotheses based on observations you made in the room, think about how you would test your hypotheses. Choose one of your hypotheses in the previous section and describe how you would test it.

Test for hypothesis #_____.

With your experimental design in place, it is a good idea to name your variables as you understand them.

What is your **independent variable**? (What you will be deliberately altering during the experiment or what is the factor that is different between experimental groups?)
What are your **standardized variables** (all factors that can vary but are kept constant during the experiment)?
What is your **dependent variable** (what is measured, the results of the experiment)?

Chapter III:
Math Review

AP Biology is much more than a high school biology class. It will cover the material you would learn in the first year college core biology courses. The biggest differences between AP Biology and a regular biology class will be the ability to extrapolate basic concepts and principles to describe and understand more complex biological processes. In many cases this will involve using more math but the basic concepts of units and conversions must be recalled. For those questions requiring more in depth mathematical analysis, four-function calculators are permitted..

Things you should know:
Scientific Notation
Understanding Graphing
Basic Statistics

Things this book will cover:
When analyzing results, it is important to understand how to take the raw data and manipulate it into a form that is interpretable and validated with statistical support. Tables, graphs, and figures are used in scientific reporting to help scientists interpret the observations collected. In this section we will practice graphing and conducting statistical analysis on a dataset.

A. The Celsius, Fahrenheit, and Kelvin temperature scales

The **Fahrenheit** (°F) non-metric temperature scale was proposed in 1724 by Gabriel Fahrenheit who was looking to improve upon Galileo's thermometer by changing from an enclosed gas to mercury. Mercury has a large uniform thermal expansion, does not adhere to the glass, and its silvery color makes it easy to read.

On his scale, he found the temperature of boiling water to be 212°. He adjusted the freezing point temperature for water to 32° so that the interval between freezing and boiling would be a rational number—180°.

Anders Celsius proposed the **Celsius** scale which was designed to have 100 degrees between the boiling point temperature of water and its freezing point temperature at standard atmospheric pressure.

The use of the Celsius or Fahrenheit scale requires using negative numbers when measuring low temperatures. This proves to be inconvenient, so in the late 1800s Lord Kelvin suggested a new temperature scale. This temperature scale is based on **absolute**

zero, the temperature at which a material has cooled to the point where it has no more heat to lose (or the point at which all molecular motion ceases). This temperature is the basis for the **Kelvin** scale and the value zero kelvin is assigned to it. At sea level, water freezes at around 273 K and boils at 373 K. Notice that there are still 100 degrees between the freezing and boiling point temperatures on the Kelvin scale. Temperatures on this scale are called **kelvins**, *not* degrees kelvin, kelvin is *not* capitalized, and the symbol (capital K) stands alone with no degree symbol.

The Celsius scale is now defined as follows:

1. The triple point of water is defined to be $0.01°$ C.
2. A degree Celsius equals the same temperature change as a degree on the ideal-gas scale, now called the **Kelvin scale**.
3. On the Celsius scale the boiling point of water at standard atmospheric pressure is $99.975°$ C in contrast to the $100°$ originally defined by the Centigrade scale.

Conversions between scales are simple but require a little math:

- To convert from Celsius to Fahrenheit, multiply by 1.8 and add 32:
$$°F = 1.8 \times °C + 32$$
From Fahrenheit to Celsius, subtract 32 and divide by 1.8:
$$°C = (°F - 32) / 1.8$$

- To convert from Celsius to Kelvin, add 273 (273.15 to be more exact) to the Celsius temperature: $K = °C + 273$
From Kelvin to Celsius, subtract 273 from the Kelvin temperature:
$$°C = K - 273$$

Common temperature comparisons

temperature	degree Celsius	degree Fahrenheit
symbol	**°C**	**°F**
boiling point of water	100.0	212.0
average human body temperature	37.0	98.6
average room temperature	20.0 to 25.0	68.0 to 77.0
melting point of ice	0.0	32.0

B. Units

There are a number of important physical quantities that have unique units that are important to remember for conversions. Many of these will be given as part of the test. However, it might be necessary to convert between different units or to derive one unit from another when making calculations for a specific question.

Derived units measure a quantity that may be expressed in terms of other units. The derived units important for chemistry are:

Derived quantity	Unit name	Expression in terms of other units	Symbol (if any)
Length	**meter**	**m**	
Area	square meter	m^2	
Volume	cubic meter	m^3	
	liter	$dm^3 = 10^{-3}\ m^3$	L or l
Mass	unified atomic mass unit	$(6.022 \times 10^{23})^{-1}\ g$	u or Da
Time	minute	60 s	min
	hour	60 min = 3600 s	h
	day	24 h = 86400 s	d
Speed	meter per second	m/s	
Acceleration	meter per second squared	m/s^2	
Temperature*	degrees Celsius	K-273.15°	°C
Mass density	gram per liter	$g/L = 1\ kg/m^3$	
Amount-of-substance concentration (molarity†)	molar	mol/L	M
Molality‡	molal	mol/kg	m
Chemical reaction rate	molar per second†	M/s = mol/(L•s)	
Force	newton	$m•kg/s^2$	N
Pressure	pascal	$N/m^2 = kg/(m•s^2)$	Pa
	standard atmosphere	101325 Pa	atm
Energy, Work, Heat	joule	$N•m = m^3•Pa = m^2•kg/s^2$	J
	nutritional calorie	4184 J	Cal
Heat (molar)	joule per mole	J/mol	
Heat capacity, entropy	joule per kelvin	J/K	
Heat capacity (molar), Entropy (molar)	joule per mole kelvin	J/(mol•K)	
Specific heat	joule per kilogram kelvin	J/(kg•K)	
Power	watt	J/s	W
Electric charge	coulomb	s•A	C
Electric potential, electromotive force	volt	W/A	V
Viscosity	pascal second	Pa•s	
Surface tension	newton per meter	N/m	

*Temperature differences in kelvin are the same as in degrees Celsius. To obtain degrees Celsius from Kelvin, subtract 273.15 (see below).

†Molarity is considered to be an obsolete unit by some physicists.

‡**Molality, m, is often considered obsolete. Differentiate m and meters (m) by context.**

These are commonly used non-SI units.

Decimal multiples of SI units are formed by attaching a **prefix** directly before the unit and a symbol prefix directly before the unit symbol. SI prefixes range from 10^{-24} to 10^{24}. Only the prefixes you are likely to encounter in chemistry are shown below:

Factor	Prefix	Symbol	Factor	Prefix	Symbol
10^9	*giga-*	G	10^{-1}	*deci-*	d
10^6	*mega-*	M	10^{-2}	*centi-*	c
10^3	*kilo-*	k	10^{-3}	*milli-*	m
10^2	*hecto-*	h	10^{-6}	*micro-*	μ
10^1	*deca-*	da	10^{-9}	*nano-*	n
			10^{-12}	*pico-*	p

Example 1: 0.0000004355 meters is 4.355×10^{-7} m or 435.5×10^{-9} m. This length is also 435.5 nm or 435.5 nanometers.

Example 2: Find a unit to express the volume of a cubic crystal that is 0.2 mm on each side so that the number before the unit is between 1 and 1000.

Solution: Volume is length × width × height, so this volume is $(0.0002 \text{ m})^3$ or $8 \times 10^{-12} \text{ m}^3$. Conversions of volumes and areas using powers of units of length must take the power into account. Multiply the factor in the chart above by the power of the unit to obtain the new exponent, as follows:

$$1\text{m}^3 = 10^3\,\text{dm}^3 = 10^6\,\text{cm}^3 = 10^9\,\text{mm}^3 = 10^{18}\,\mu\text{m}^3$$

The length 0.0002 m is 2×10^2 μm, so the volume is also 8×10^6 μm³. This volume could also be expressed as 8×10^{-3} mm³, but none of these numbers are between 1 and 1000.

Expressing the volume in liters is helpful in cases like these. There is no power on the unit of liters, therefore:

$$1\text{L} = 10^3\,\text{mL} = 10^6\,\mu\text{L} = 10^9\,\text{nL}$$

Converting cubic meters to liters gives:

$$8 \times 10^{-12}\,\text{m}^3 \times \frac{10^3\,\text{L}}{1\text{m}^3} = 8 \times 10^{-9}\,\text{L}$$

The crystal's volume is 8 nanoliters (8 nL).

Dimensional analysis is a structured way to convert units. It involves a conversion factor that allows the units to be canceled out when multiplied or divided. A **conversion factor** is the same measurement written as an equivalency between two different units such as 1 meter = 100 centimeters.

One Dimension Unit Conversions

Example 3: Convert 6.0 cm to km.

Solution:

Since 1cm = (1/100) m or 10^{-2} m and 1m = 1/1000 km or 1.0×10^{-3} km,

$6\text{cm} = 6 \times 10^{-2}$ m = 6.0×10^{-5} km.

=>therefore, $6\text{cm} = 6.0 \times 10^{-5}$ km

C. Graphing

It is sometimes difficult to recognize patterns or trends in the raw data that scientists collect; instead, scientists frequently use graphs to organize and visualize their findings. The three most common types of graphs are pie graphs, bar graphs, and line graphs. **Pie graphs** are used to represent proportional data, **bar graphs** are used for distinct classes of data, and **line graphs** are used to represent progressive series of data.

All types of graphs should include the following:

- A descriptive title.
- An X-axis and a Y-axis with an appropriate range of units and equal intervals unless otherwise indicated.
    - o The X-axis (horizontal) contains the scale for the independent variable.
    - o The Y-axis (vertical) contains the scale for the dependent variable.
- The axes labeled with name of the variable and its appropriate units.
- Graph type (line, bar, or pie) that is appropriate for the data type.
- A key if more than one set of data is presented on one graph.

Example

Here we have an example set of data for you to graph. These experimental results were collected using cockroaches, feeding them different concentrations of pesticides, and measuring their response time to run into a shelter at a fixed distance when the lights are turned on in the lab.

The effect of ingested pesticide on shelter-seeking performance in cockroaches.

Animal ID	Age (days)	Time in dark (hours)	Pesticide ingested (ul/mg)	Time to shelter (seconds)
AB232	9	2	0	1.8
CH321	9	2	0	2.0
AB134	9	2	0	1.6
AB368	9	2	0.15	3.1
CH827	9	2	0.15	3.4
CH622	9	2	0.15	4.7
CH138	9	2	0.35	10.1
AB268	9	2	0.35	10.7
AB116	9	2	0.35	14.5
CH062	9	2	0.55	13.7
CH526	9	2	0.55	18.2
AB024	9	2	0.55	16.4

Before you begin graphing the data, there are a few questions you should ask about the table.

- What is the hypothesis being tested in this experiment?
- Which column in the table is the independent variable? The dependent variable? The standardized variables?

- Which rows are the control animals and which are the experimental? How do you know? How many experimental groups were tested?

Following the rules for graphing, and the experimental results in the table above, graph the data in a line graph.

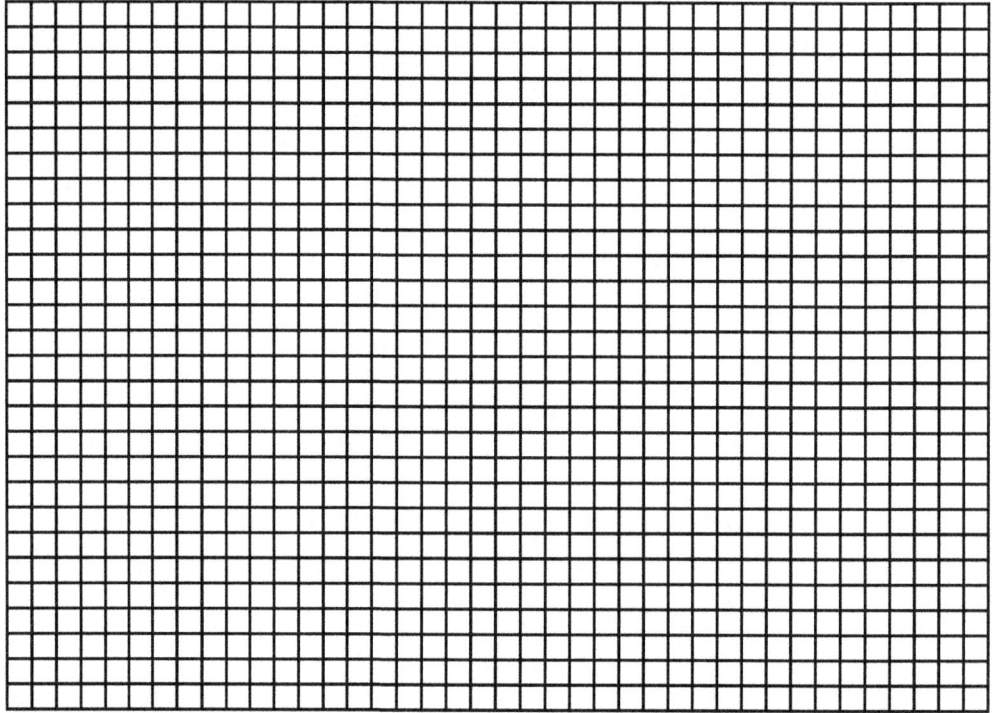

D. Statistics

Statistics serve as an essential tool for analyzing experimental results. Statistical tests employ mathematical formulas that allow scientists to calculate the probability whether the trend observed in their experiments has occurred beyond a simple coincidence.

In this section we will take real data and practice a statistical test called the chi-squared test (χ^2-test). A **chi-square test** is a statistical test to compare the observed data with the data we would expect to obtain according to our hypothesis. It can then be used to reject the hypothesis that the data are independent.

Example

We would expect that a tossed coin would have equal chances of landing on either side, so the probability is 0.5 of a single coin flip to be either heads or tails. We must first establish our null hypothesis, that the coin is 'fair' and will be equally likely to land on "heads" or "tails." For 200 tosses, we would expect 100 heads and 100 tails.

In the example below, we have data collected from flipping a coin 200 times and documenting the number of "heads" or "tails."

Extra credit: if you would like, you can collect the data yourself and replace the values in the table below in the observed row of heads and tails.

Results of the Coin Flip Experiment

	Heads	Tails	Total
Observed	111	89	200
Expected	100	100	200
Total	211	189	400

The 'Observed' values are those we gather ourselves. The 'Expected' values are the frequencies expected, based on our null hypothesis. We total the rows and columns as indicated. You can see that we do not have an equal number of heads and tails observed and therefore we need to determine if the observed deviates from what is expected or is within probable coincidence. Using probability theory, statisticians have devised a way to determine if the frequency of the observed (obs.) distribution differs from the expected (exp.) distribution.

Terms and symbols:

Mean = sum of all data points divided by number of data points

Median = middle value that separates the greater and lesser halves of a data set

Mode = value that occurs most frequently in a dataset

Range = value obtained by subtracting the smallest observation from the greatest

n = size of the sample

obs. = observed individuals with observed genotype

exp. = expected individuals with observed genotype

df = degrees of freedom refers to the number of values in the final calculation that are free to vary, = n – 1

Mean/average

$$x = \text{mean}$$

$$x = \frac{1}{n} \sum_{i=1}^{n} x_i$$

Standard deviation

$$s = \text{sample standard deviation}$$

$$s = \sqrt{\frac{\sum (x_i - x)^2}{n-1}}$$

Standard error

$$SE = \frac{s}{\sqrt{n}}$$

The chi-square test:

$$x^2 = \sum \frac{(\text{Observed value} - \text{Expected value})}{\text{Expected value}}$$

In our example, we have two classes, "heads" and "tails" so we must calculate chi-square the following way:

$$x^2 = \frac{(\text{obs. heads} - \text{exp. heads})^2}{\text{exp.}} + \frac{(\text{obs. tails} - \text{exp. tails})^2}{\text{exp.}}$$

Now plug in the values:

$$x^2 = \frac{(111-100)^2}{100} + \frac{(89-100)^2}{100}$$

$$x^2 = \frac{(11)^2}{100} + \frac{(-11)^2}{100}$$

$$x^2 = 1.21 + 1.21$$

$$x^2 = 2.42$$

Now we have to consult a table of critical values of the chi-squared distribution. Here is a portion of such a table:

Chi-square critical values table

df/ prob.	0.99	0.95	0.90	0.80	0.70	0.50	0.30	0.20	0.10	0.05
1	0.00013	0.0039	0.016	0.64	0.15	0.46	1.07	1.64	2.71	3.84
2	0.02	0.10	0.21	0.45	0.71	1.39	2.41	3.22	4.60	5.99
3	0.12	0.35	0.58	1.00	1.42	2.37	3.66	4.64	6.25	7.82
4	0.3	0.71	1.06	1.65	2.20	3.36	4.88	5.99	7.78	9.49

The left-most column lists the degrees of freedom (df). We determine the degrees of freedom by subtracting one from the number of classes. In this example, we have two classes, "heads" and "tails." So we take two classes minus one and we arrive at a df = 1. Our chi-squared value is 2.42. Move across the row for 1 df until we find critical numbers in the range of our value. In this case, 1.64 and 2.71. If we look at the top of the column for these two numbers we see they are corresponding to a probability of 0.20 and 0.10 respectively when we follow along the row of the table. We can interpolate our value of 2.42 to estimate a probability of 0.12. This value means that there is an 88% chance that our coin is biased. In other words, the probability of getting 111 heads out of 200 coin tosses with a fair coin is 12%. In biology we often accept 5% or less to be significant. Because the chi-squared value we obtained in the coin example is greater than 0.05, we accept the null hypothesis as true and conclude that our coin is fair.

How to Answer the Grid-In Response on the AP Biology Exam

In the 90 minutes available for Section I there are 63 multiple-choice questions and 6 grid-in questions. These grid-in questions are few but can be complex. By remembering these seven key tips you will be well prepared for tackling these six questions.

1. Read the diagram. You have to know what you are looking for, first you must carefully read and interpret the diagrams provided.

2. Keep track of your units. You do not have to provide units in your answers on the grid-in questions; however, looking at the units may give you a hint as to what you are being asked to find.
3. **Read and reread** the directions carefully.
4. **Do not** round until the very end.
5. There is **no** exponent function on the calculator. So practice before you take the test.

Example: $4^3 = 4 \times 4 \times 4 = 64$

6. Be able to convert between scientific notation and whole numbers.

Examples: $3.21 \times 10^4 = 32100$

$3.21 \times 10^{-4} = 0.000321$

Example of the grid-in chart

Answers can start in any column and any extra columns should be left blank.

Use decimals and symbols when needed.

Only one circle per column.

SECTION II:
Content Review

Chapter 1.
Big Idea 1: Evolution

Big Idea 1: Evolution: The process of evolution drives the diversity and unity of life

What you will learn from this chapter:

- The process of natural selection
- Contributions of phenotypic variation and genetic drift
- Multi-disciplinary understanding of evolution
- Phylogenetic trees and cladograms
- Evidence for speciation and extinction
- Hypotheses for the origin of living systems

Introduction

The development of life on Earth, from a simple to more complex form, is the foundation on which the study of biology is built. In this chapter, we will discuss what affects the genetic makeup of populations, and how the variation in individuals leads to increased survival and therefore, the increased chance of passing these favorable traits to future generations. You will see how evidence from many scientific disciplines, including mathematics, supports the idea that genetic makeup contributes to understanding our history and common ancestry. Let's begin!

Enduring Understanding 1.A. Change in the genetic makeup of a population over time is evolution

1.A.1: Natural selection is a major mechanism of evolution.

Natural selection is one of the basic mechanisms of evolution, and is often referred to by its idea of: "survival of the fittest."

Darwin's grand idea of evolution by natural selection is relatively simple but often misunderstood. To find out how it works, imagine a population of beetles.

1. There is variation in traits.
For example, some beetles are green and some are brown.
2. There is differential reproduction.
    Since the environment can't support unlimited population growth, not all individuals get to reproduce to their full potential. In this example, green beetles tend to get eaten by birds and reproduce less often than brown beetles do.
3. There is heredity.
    The surviving brown beetles have brown baby beetles because this trait has a genetic basis that is passed down the next generation.
4. End result:
    The more advantageous trait, brown coloration, which allows the beetle to have more offspring, becomes more common in the population. If this process continues, eventually, all individuals in the population will be brown.

Allele frequency, or **gene frequency,** is the relative frequency of an allele (variant of a gene) at a particular locus in a population, expressed as a fraction or percentage. Specifically, it is the fraction of all chromosomes in the population that carry that allele.

Given the following:

- a particular locus on a chromosome and a given allele at that locus
- a population of N individuals with ploidy n, i.e. an individual carries n copies of each chromosome in their somatic cells (e.g. two chromosomes in the cells of diploid species)
- the allele exists in i chromosomes in the population
- then the allele frequency is the fraction of all the occurrences i of that allele and the total number of chromosome copies across the population, $i/(nN)$.

The allele frequency is distinct from the genotype frequency, although they are related, and allele frequencies can be calculated from genotype frequencies.

In population genetics, allele frequencies are used to describe the amount of variation at a particular locus or across multiple loci. When considering the ensemble of allele frequencies for a large number of distinct loci, their distribution is called the allele frequency spectrum.

If (pp), (qq), and (pq) are the frequencies of the three genotypes at a locus with two alleles, then the frequency p of the A-allele and the frequency q of the B-allele in the population are obtained by counting alleles.

$$p = (pp) + 1/2\ (pq) = \text{frequency of A}$$
$$q = (pq) + 1/2\ (qq) = \text{frequency of B}$$

Because p and q are the frequencies of the only two alleles present at that locus, they must sum to 1. To check this:

$$p+q = f(AA) + f(BB) + f(AB) = 1$$
$$p = 1-q \text{ and } q = 1-p$$

If there are more than two different allelic forms, the frequency for each allele is simply the frequency of its homozygote plus half the sum of the frequencies for all the heterozygotes in which it appears.

These explanations for gene frequency are commonly associated with the Hardy-Weinberg law which states that frequencies of genotypes will be in equilibrium if the conditions adhere to these five factors: a large population, no mutations, no migration, random mating, and no natural selection. As a result, the equation to determine the frequency of the genotypes of a population is:

$$p^2 + 2pq + q^2 = 1$$

1.A.2: Natural selection acts on phenotypic variations in populations.

A **population** is composed of individuals with different genetic backgrounds that result in different phenotypes. Height, shoe size, eye color, and hair color are examples of variable phenotypes of human populations. But not all variations of a trait are beneficial, or efficient, under all natural conditions. Under certain conditions, individuals with a set of advantageous traits will produce more offspring, i.e. will have a higher **fitness**. Through heredity, the offspring will also show the advantageous traits, or new combinations of them, which will in turn be passed along. With time, the underlying genotypes responsible for such phenotypes increase in frequency in the population. However, changes in the

The environment is always changing, there is no "perfect" genome, and a diverse gene pool is important for the long-term survival of a species.

Phenotypic variations are not directed by the environment but occur through random changes in the DNA and through new gene combinations.

environment, natural or anthropogenic, may reverse the situation and favor different phenotypes. As long as there's diversity, there's room for **natural selection** to occur.

Natural selection is easily illustrated with the case of the peppered moth (*Biston betularia*). This moth occurs in the temperate region, and typically shows a light-colored body. Around the industrial revolution, scientists started noticing that a dark-colored variation of this species was becoming more and more frequent in heavily polluted cities. Much later, after Darwin and the publication of his famous *The Origin of Species*, it was postulated that such variation in color could be evidence of natural selection. Experiments performed in the 1950's showed that, in unpolluted cities, light-colored moths could successfully camouflage in trees, escaping from predation by birds. In polluted cities, trees were darkened by soot, making the light-colored varieties more conspicuous to predators. Heavy predation decreases the chance of producing offspring, so with time the dark-colored phenotype became more abundant in the areas where it had a higher chance of surviving due to an effective camouflage.

Natural selection may favor different phenotypes if conditions change. **Sickle cell anemia** is a disease caused by a single mutation in the hemoglobin gene. Heterozygous individuals live relatively normal lives, as one copy of the gene still produces the normal hemoglobin for the blood cells. Homozygous individuals for this condition have reduced life expectancy, and their frequency in a population tends to decrease. However, in areas where malaria (caused by a species of the protozoan *Plasmodium*) is endemic, heterozygous individuals have a lower risk of dying of malaria, due to a protective effect caused by the abnormal red blood cells. In such areas, having one copy of the mutation is therefore advantageous, and the frequency of heterozygous individuals tends to increase.

Resistance to antibiotics is also a result of natural selection, and can be observed over days or weeks. The misuse of antibiotics is one of the main causes of drug resistance. For instance, during a treatment with antibiotics, susceptible bacteria are eliminated first. Cells carrying a mutation, or one or more traits conferring resistance, will survive longer. If the treatment is incomplete, or not long enough to effectively eliminate all cells, the remaining cells which carry the resistant traits will reproduce and proliferate, producing an entire strain of resistant cells. Examples of drug-resistant bacteria, responsible for many infections in hospitals, are the methicillin-resistant *Staphylococcus aureus* (MRSA), vancomycin-resistant *S. aureus* (VRSA), and the vancomycin-resistant *Enterococcus* (VRE). Besides bacteria, fungi, protozoans, viruses, and insects can also develop drug resistance.

Even without knowing, humans have been playing with natural selection for centuries. By observing plants and animals with desirable traits, farmers would select and cross them with other individuals showing another set of favorable characteristics, thus creating a new breed (animals), or cultivar (plants). This is known as **artificial selection**, or artificial breeding. The rice, wheat, and maize we eat today are very different from the original wild type species. In artificial selection, inbreeding is sometimes used to reinforce the desirable traits. Many modern-day animal races, as well as the domestication of animals, are the product of artificial selection.

Extensive artificial selection, such as the methods used to produce crops with more favorable taste or growth pattern, has a down side. To maximize productivity in agricultural monocultures, extensive areas are seeded with mass-produced and/or cloned

Some phenotypic variations significantly increase or decrease fitness of the organism and the population.

seeds, meaning that genetic diversity is low or nonexistent. As seen above, natural selection can only act on phenotypic diversity. Thus, upon environmental changes, such as climate change, or the introduction of a new pest, all individuals would be susceptible and it is possible that the whole monoculture would be lost.

1.A.3: Evolutionary change is also driven by random processes.

Natural selection can be viewed as a directional process that selects advantageous traits. However, evolution also operates through **neutral processes,** where one or more traits are fixed in the populations due to chance. **Genetic drift** is the random change in the frequency of alleles from one generation to the next. Over time, these changes tend to eliminate alleles with few copies in the population, decreasing the genetic diversity. The **founder effect** also leads to the loss of genetic diversity. This occurs when a new population is established from a small pool of organisms. All descendants, the future generations of that population, will have a combination of a fewer alleles, compared to the original, more diverse, donor population. If the new population is isolated, and there's no immigration from adjacent populations to ensure the entrance of new alleles, the founder effect can lead to speciation. A **bottleneck** event occurs when the size of a population is suddenly and drastically reduced, eliminating many alleles at once. Bottleneck events can be natural, such as due to floods and fires, or due to human activities. If, by any chance, the remaining individuals were not the fittest, the population could disappear. In all cases, the genetic diversity in the populations affected by natural processes will increase over time due to mutations.

Genetic drift is a nonselective process occurring in small populations.

1.A.4: Biological evolution is supported by scientific evidence from many disciplines, including mathematics.

One of the key factors that supports evolutionary theory is the use of fossil records. Fossils provide glimpses into the biological history of the Earth. Geologists and paleontologists use scientific principles in chemistry and physics to estimate the age of rock formations. Fossils found in rock layers help paleo-biologists observe physical changes in structures of organisms, identify common ancestors of living and extinct animals, evaluate changes in biological diversity and hypothesize how environments have changed throughout Earth's history. One issue with fossil records is that they provide a limited source of material since only organisms with tissues that can be mineralized can be observed therefore, fossils represent a very small percentage of the organisms that ever existed.

Reduced genetic variation within a given population can increase the differences between populations of the same species.

The ability to compare anatomical structures and character traits of organisms is another way of estimating their evolutionary relationships. Organisms that have similar traits could likely have had a common ancestor or those traits would have evolved independently. **Homologous structures** are shared physical traits of organisms that had a common ancestor with a similar trait. For example, the forelimb bones of most vertebrates are a homologous structure (see figure below). Though they may not have the same function, they have the same pattern of bone numbers and locations. Alternatively, the **analogous structures** share a similar function but do not have a common ancestor. An example of this would be the wings of a moth that allow it to fly like the wings of a bat or bird, but they evolved independently and, through convergent evolution, share the function of flight.

Molecular, morphological and genetic information of existing and extinct organisms add to our understanding of evolution.

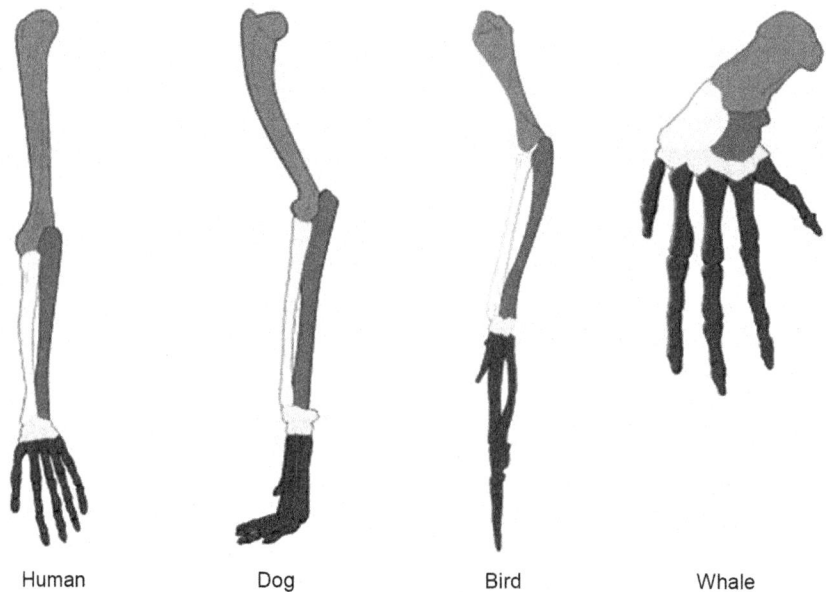

Figure depicting homology between vertebrate forelimb structure.

We can use the observed information about the character traits of organisms and the genes that encode them to determine their likely relationship. In other words, we can use the genetic differences between organisms to draw a diagram of their expected evolutionary relationship. By understanding mutation rates, genetic drift, and natural selection, we can determine the relationship between organisms as well as their inferred common ancestors.

Enduring Understanding 1.B: Organisms are linked by lines of descent from common ancestry

1.B.1 Organisms share many conserved core processes and features that evolved and are widely distributed among organisms today.

Structural and functional evidence supports the relatedness of all domains.

The oldest record of life on Earth is from about 3.5 billion years ago, while the oldest record of eukaryotic life dates from about 1.8 billion years ago. This indicates that prokaryotes were the only life form during most of the Earth's history. Photosynthetic life forms appeared around 3 billion years ago and slowly started filling the atmosphere with oxygen, which later allowed the great diversification of aerobic eukaryotes. The **Endosymbiotic Theory (or Symbiogenesis Theory)** proposes that eukaryotes arose from symbiotic groups of prokaryotic cells. According to this theory, smaller prokaryotes lived within larger prokaryotic cells, eventually evolving into chloroplasts and mitochondria, thereby originating membrane-bound organelles seen in modern day eukaryotic cells. Chloroplasts are the descendants of photosynthetic prokaryotes, and mitochondria are likely the descendants of bacteria that were aerobic heterotrophs. This theory is supported by the many similarities observed between bacteria and mitochondria and chloroplasts (see diagram below).

New mitochondria and plastids are formed in a similar way to binary fission	The organelles' ribosomes are 70S, as in bacteria, while cytoplasmic ribosomes are 80S
Similarities between bacteria and mitochondria/chloroplasts	
They have a single circular DNA molecule	Some chloroplasts have a peptidoglycan cell wall in the membrane

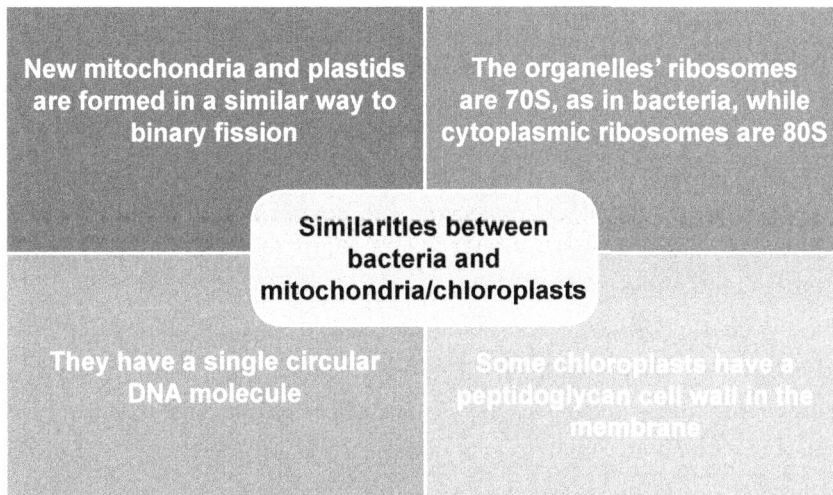

1.B.2: Phylogenetic trees and cladograms are graphical representations of evolutionary history that can be tested.

Phylogenetic trees and **cladograms** are diagrams that depict inferred evolutionary relationships among organisms. Cladograms compare character traits while phylogenetic trees are cladograms that reference genetic information comparing similarities and the differences among species. Scientists use this information to better understand the origin of traits and infer explanations of their origins. The pentadactyl limb of ancestral vertebrates is an example of a basic characteristic that can be visualized in a cladogram (see figure below). It is interesting to note that homologous structures do not always predict phylogenetic relationships.

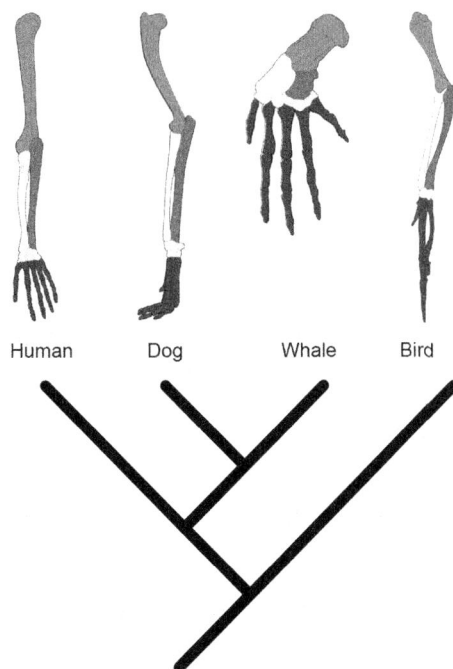

Human   Dog   Whale   Bird

The phylogenetic relationship between forelimbs of humans, dogs, whales, and birds.

Phylogenetic trees and cladograms illustrate speciation that has occurred, in that relatedness of any two groups on the tree is shown by how recently two groups had a common ancestor.

Phylogenetic trees and cladograms can be constructed from morphological similarities of living or fossil species, and from DNA and protein sequence similarities, by employing computer programs that have sophisticated ways of measuring and representing relatedness among organisms.

Enduring Understanding 1.C: Life continues to evolve within a changing environment

1.C.1: Speciation and extinction have occurred throughout the Earth's history.

Speciation rates can vary, especially when adaptive radiation occurs when new habitats become available.

It is estimated that 99.9999% of all species that have ever lived are extinct. Extinction has always been a major force in macroevolution. Species do not last forever; the mean expected lifespan of a marine bivalve species is about 14 million years, and the mean expected lifespan of a terrestrial mammal species is about a tenth of that – 1.4 million years. The expected life span of a species is considered the **extinction rate**. Climate change, natural disasters, and other phenomena have always caused extinction and even large scale **extinction events** (think, the dinosaurs). As some species originate, they inevitably drive other species to extinction. Extinctions, in turn, pave the way for speciation. **Speciation** is an event where two organisms that once shared a common ancestor are now biologically distinct and reproductively isolated from each other. An **adaptive radiation** is a wave of speciation that occurs as a new habitat is colonized by a lineage, or in the wake of the extinction of another lineage. An adaptive radiation of mammals followed the extinction of the dinosaurs.

Species extinction rates are rapid at times of ecological stress.

One factor in modern extinction events is the **human impact on ecosystems**. Since the development of agriculture 10,000 years ago, humans have modified an increasing proportion of the Earth's resources for our own purposes. The impact of humans has caused extinction rates to be 10 to 1000 times greater than any time in the last 100,000 years. For example, one estimate for the recent background extinction rate for birds is one species extinction per 400 years. If only this natural rate of loss affected the number of bird species, no more than two extinctions should have occurred in the past 800 years. Scientists estimate that the actual loss during this time period lies somewhere between 200 and 2,000 species. We have set in motion a mass extinction, one of the largest, that will not culminate until thousands of years from now. The major causes of anthropogenic extinction are linked to habitat destruction and habitat fragmentation, habitat change and disruption of ecosystem processes, introduction of exotics, and overexploitation due to hunting and poaching.

1.C.2: Speciation may occur when two populations become reproductively isolated from each other.

Speciation results in diversity of life forms. Species can be physically separated or various pre-and post-zygotic mechanisms can maintain reproductive isolation and prevent gene flow.

Reproductive isolation is one of the driving explanations behind evolutionary theory. Darwin himself observed this in finches and tortoises on the Galapagos Islands that had clear similarities to those species found on the mainland. Being geographically isolated prevented migration back and forth from the mainland species and because each new island brought new niches, the finches and tortoises had to adapt to their new environments. The mechanisms of reproductive isolation have been classified in a number of ways. Zoologist Ernst Mayr classified the mechanisms of reproductive isolation in two broad categories: those that act before fertilization (or before mating in the case of animals, which are called pre-copulatory) and those that act after fertilization. These have also been termed pre-zygotic and post-zygotic mechanisms. Each of these mechanisms of reproductive isolation are genetically controlled and it has been demonstrated that species can evolve with geographic distribution overlaps (**sympatric speciation**) or as the result of adaptive divergence that accompanies **allopatric speciation**.

One classic example of speciation includes the *Ensatina* salamanders of California. Their habitat in the mountainous regions of the area seem to have evolved into seven distinct sub-species reproductively isolated from each other. They are isolated by geographical barriers mountains and valleys – so that each sub-species has adapted their size, coloration, and body markings to their local environments. The populations following the inland Sierra Mountains have camouflage style markings while those populations that follow the coastal mountains have colorings that mimic those of poisonous newts.

1.C.3: Populations of organisms continue to evolve.

Organisms continue to evolve and adapt to new stressors in their environment. One excellent example is the adaptation to resist chemicals such as pesticides and antibiotics. Agriculture has benefitted greatly from the development of pesticides. We have greater food security because of them. However, with the increased yield comes increased pests such as weeds and insects. To combat these, chemicals are often applied that are toxic to the pest animal or plant. Populations of these pests are under intense pressure to adapt to the exposure of these toxic chemicals and as such, the best-adapted will pass on the genes required to resist the toxic effects. This is best exemplified by the observation of farmers in the 1950s who lost about seven percent of their crops to pests; these rates have more than doubled, reaching to 15 percent. The artificial selection on the pest populations produces individuals with "genetic shielding" for a higher tolerance. Local populations retain these new phenotypes as they improve overall reproductive success.

Enduring Understanding 1.D: The origin of living systems is explained by natural processes _____

1.D.1: There are several hypotheses about the natural origin of life on earth, each with supporting evidence.

There are two major hypotheses about the origin of life on Earth. First, the **Abiogenic Theory** is the oldest and one of the most widely supported. This posits that life began approximately 3.5 billion years ago as the result of a complex sequence of chemical reactions. Non-organic molecules undergo reactions that give rise to organic molecules that further interact with one another to give rise to the early forms of life. How exactly this process occurred is not clear but the Miller-Urey experiment (see section 1.D.2) in the 1950's provided some of the basic evidence that supports this hypothesis. It is possible that land, water and atmospheric conditions led to the spontaneous molecules of life. There are even some modern hypotheses that suggest that life could have first originated deep in the ocean surrounding volcanic vents (see section 1.D.2).

The primordial soup theory is just one of many abiogenesis theories. Another hypothesis came from the observation that clay can act as a catalyst of RNA polymerization and of formation of membranes by lipids. The **Clay hypothesis** proposes that organic molecules could have been formed in solutions of clay minerals, such as silica. The **Iron-sulfur World hypothesis** suggests that life evolved underwater, in hot and high-pressure hydrothermal vents rich in sulfides.

New species arise from reproductive isolation over time, which can involve scales of hundreds of thousands or even millions of years, or speciation can occur rapidly through mechanisms such as polyploidy in plants.

Scientific evidence supports the idea that evolution has occurred in all species, and continues to occur.

Primitive Earth provided inorganic precursors from which organic molecules could have been synthesized due to the presence of available free energy and the absence of a significant quantity of oxygen.

An alternative hypothesis is **Panspermia,** in which life, or the fundamental molecules for life, exists throughout the universe and through the movement and transport of meteorites and comets life can be seeded on planetoids. According to Panspermia, life on earth would have originated somewhere else in the universe and was brought to our planet through these astronomical movements.

1.D.2: Scientific evidence from many different disciplines supports models of the origin of life.

The **Abiogenic Theory** proposes that life developed on Earth from nonliving materials. This transformation had four stages. The first stage was the synthesis of small organic monomers, such as amino acids and nucleotides, from inorganic molecules. In the second stage, these monomers combined to form polymers, such as proteins and nucleic acids. The third stage involves formation of **protobionts**, droplets containing proteins or nucleic acids surrounded by a membrane-like structure. The last stage is the origin of heredity and genetic information. The first genetic material was probably RNA, not DNA, as it is capable of self-replication, and it can be used to synthesize DNA. This is known as the **RNA World** hypothesis.

In the 1920s, Oparin and Haldane proposed the conditions that could have favored the formation of organic molecules from simple, abundant, inorganic molecules. They suggested that the primitive atmosphere lacked oxygen but was rich in hydrogen, methane, water vapor, and ammonia. Under such reducing conditions, and with the input of energy from lightning and UV radiation, these molecules combined to form organic molecules in Earth's primitive oceans. This theory is known as the **Primordial Soup** theory of the origin of life. Later, in the 1950s Miller and Urey designed a laboratory experiment to test Oparin and Haldane's theory by introducing electrical sparks into a reducing system containing those molecules, and demonstrating the ability to synthesize simple amino acids and fatty acids (see figure below).

The RNA World hypothesis proposes that RNA could have been the earliest genetic material.

Geological evidence provides support for models of the origin of life on Earth.

Molecular and genetic evidence from extant and extinct organisms indicates that all organisms on Earth share a common ancestral origin of life.

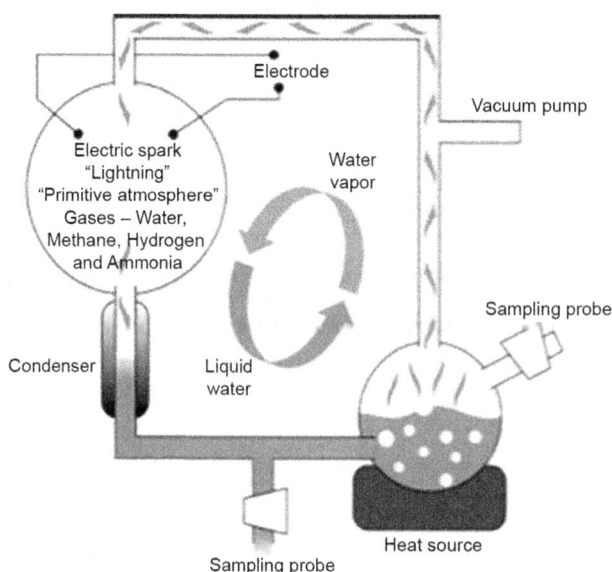

The **Panspermia** hypotheses have had a growing movement of support due to recent findings: the collection and observation of interstellar dust containing many organic molecules that have the potential, if met with ideal conditions on a new planet's surface, for abiogenic life to begin. Similarly, if life had evolved on another planet, it could have jettisoned from that planet through impacts of meteoroids carrying microbial hitchhikers. Experiments on the International Space Station observed both prokaryotic and eukaryotic phototrophs surviving for 548 days in the vacuum of space.

Keywords

Allele frequency
    Hardy- Weinberg law
Population
Fitness
Natural selection
    Neutral processes
    Genetic drift
    Founder effect
    Bottleneck event
Artificial selection
Homologous/analogous structures
Endosymbiotic/Symbiogenesis Theory
Phylogenetic tree
Cladogram
Extinction rate
Speciation
    Sympatric speciation
    Allopatric speciation
Miller-Urey experiment
Abiogenic theory
    Primordial soup
    Protobionts
    RNA World hypothesis
    Clay hypothesis
    Iron-sulfur World hypothesis
Panspermia

Summary

The process of evolution is defined as a change in the genetic makeup of a population over time, and the major force behind evolution is natural selection—in which favorable phenotypes are more likely to survive and produce more offspring.

Catastrophic events of nature, random environmental changes, and human-induced events can all confer changes in the gene pools of populations.

The diversity of life is explained by the fact that speciation and extinction have continually occurred, and life continues to evolve, within a changing environment on Earth.

Conserved core processes and genetic information are shared by Archaea, Bacteria, and Eukarya, and this provides evidence that all organisms are linked from a common ancestry.

There are different hypotheses on the origin of life, however, experiments have demonstrated that chemical and physical processes can give rise to more complex molecules. The first genetic material may have been RNA and aided in the evolution of bacteria, which eventually gave rise to eukaryotic cells.

Chapter 1 Quiz _____

1. **Which of the following represents the most likely timeline of the evolution of life on Earth, starting with the oldest?**

   A. Plants → fungi → animals → humans

   B. Fungi → bacteria → plants → protists

   C. Bacteria → protists → plants → animals

   D. Protists → humans → animals → plants

Questions 2–6 relate to the phylogenetic tree that traces the evolution of plants.

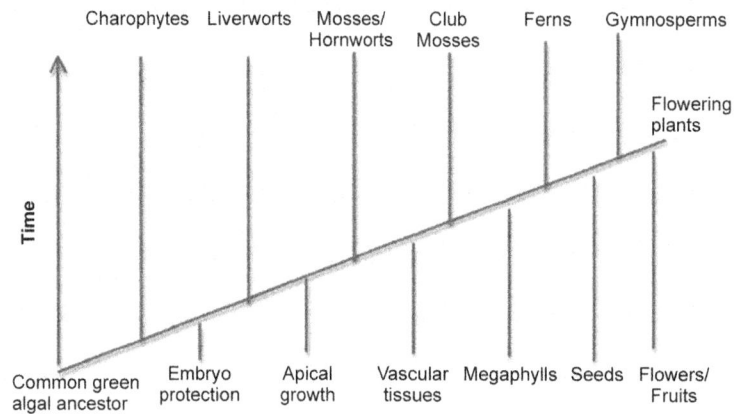

2. **Which trait was the earliest to appear in the evolution of plants?**

   A. Seeds

   B. Flowers

   C. Apical growth

   D. Embryo protection

3. **Which two types of plants are most closely related to each other?**

   A. Hornworts and ferns

   B. Flowering plants and gymnosperms

   C. Charophytes and club mosses

   D. Liverworts and flowering plants

4. **Which plants never evolved the ability to have apical growth?**

   A. Ferns

   B. Charophytes

   C. Hornworts

   D. Gymnosperms

5. **Which structure has led to the success of the most recent land plants?**

   A. Fruits

   B. Seeds

   C. Megaphylls

   D. Vascular protection

6. **What conclusion can be drawn from the information presented in this phylogenetic tree?**

   A. Vascular tissues were essential to the success of all land plants.

   B. Ferns are the most closely related land plants to the ancient green algal ancestor.

   C. The progression of time has caused some land plants to become less complex than their ancestors.

   D. Liverworts have adapted and survived without the development of seeds or vascular tissue.

7. **The wing of a bat, the human hand, the fin of a whale, and the front leg of a dog all share a similar bone structure. How do these traits support the theory of evolution?**

   A. They show diversity between species.

   B. They show a common ancestor among all species.

   C. The show the existence of a unified anatomical theme of all organisms.

   D. They show how adaptive radiation can lead to the formation of new traits.

8. The fossil record can be used to estimate the total number of taxonomic families that have existed over time. These estimates are shown in the graph. The letters A-D represent periods of mass extinctions. What can be concluded from the data?

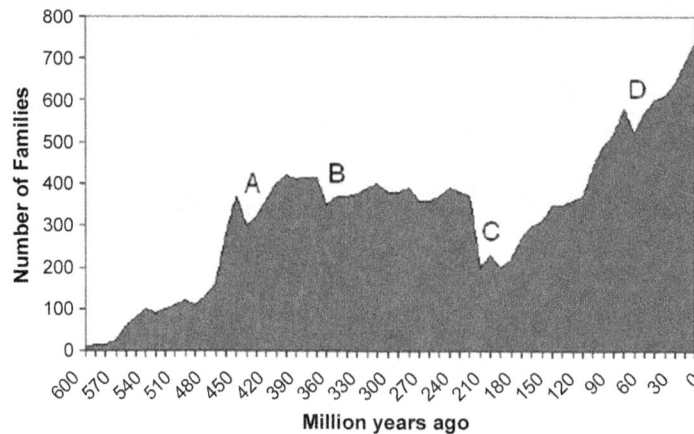

A. Changes in global climate can result in a mass extinction.

B. Mass extinctions are often followed by periods of speciation and increased biodiversity.

C. Mass extinctions generally eliminate all life on the planet, so ecosystems need to start over.

D. There is incomplete evidence from the fossil record to associate mass extinctions with the biodiversity.

9. The formation of the Grand Canyon affected land animals like squirrels differently from animals like birds because of the barrier limiting the interactions between animals. Based on this information, which of the following is the MOST CORRECT statement?

A. The squirrel population underwent allopatric speciation.

B. The bird population underwent allopatric speciation.

C. The squirrel population underwent allopatric speciation, and the bird population was not affected.

D. No speciation occurred in squirrels or birds.

10. A large sailboat with a group of 200 explorers containing a small group of people with blue eyes (an autosomal recessive trait) discovers an uninhabited island. If the explorers stay on the island, what is the most likely effect on the number of blue-eyed inhabitants after several generations?

A. There will be no genetic drift.

B. The number of blue-eyed people will decrease.

C. The number of blue-eyed people will increase.

D. The number of brown-eyed people will increase.

Chapter 1 Quiz Answer Key _____

1. **Answer: C.** The first organisms on Earth are believed to have been bacteria. These single-celled organisms specialized, became more complex, and evolved into protists. Protists are either plant-like or animal-like. The plant-like protists migrated onto land and became plants. The animal-like protists evolved into animals.

2. **Answer: D.** Plants developed a mechanism to protect their embryos early in their evolution. This showed to be successful, especially when they migrated onto land.

3. **Answer: B.** The flowering plants and gymnosperms are closest together on the phylogenetic tree. This indicates that their evolution is more closely related than that of other types of plants.

4. **Answer: B.** The charophytes that evolved soon after plants moved onto land are the earliest land plants in their evolutionary history. They never developed apical growth and are most closely related to the ancient green algal ancestor.

5. **Answer: A.** The most recently evolved plants are the flowering plants. These organisms have developed a structure called a fruit that protects and nourishes the seeds. Fruits are also used to disperse the seeds to new locations, which helps pass on the plant's genes.

6. **Answer: D.** The phylogenetic tree shows that liverworts evolved well before the development of vascular tissues. Since liverworts are still in existence, that indicates that even without these specialized tissues, they are well adapted for their particular environment.

7. **Answer: B.** The commonalities are called homologous structures. While the adult forms of birds, whales, dogs, and humans all look different, they are all vertebrates with similar basic anatomy.

8. **Answer: B.** The graph shows that after a period of mass extinction, there is a period of growth. All four of the major events show this pattern, albeit in differing intensities. The smallest period of speciation was after extinction B and the largest was after extinction C.

9. **Answer: C.** Since the squirrels could not cross the geographic barrier created by the formation of the Grand Canyon, the squirrel population underwent allopatric speciation resulting in two different species on the north and south rims of the Grand Canyon. This did not occur in the bird population because the birds could still fly and cross the barrier to interbreed, therefore they remained as one population.

10. **Answer: C.** Over many generations, it is likely that the proportion of blue-eyed people in the population will increase due to genetic drift.

Chapter 2.
Big Idea 2: Energy

Big Idea 2: Energy: Biological systems utilize free energy and molecular building blocks to grow, to reproduce, and to maintain dynamic homeostasis.

What you will learn from this chapter:

- Laws of thermodynamics
- Entropy and enthalpy
- Cellular respiration
- Electron transport chain
- Photosynthesis
- Selective permeability
- Passive and active transport
- Biotic and abiotic factors
- Dynamic homeostasis
- Non-specific immune responses
- Physiological changes due to environmental stimuli and molecular signals

Introduction

How do cells, populations, and ecosystems use and re-use energy? In this section, we will discuss bioenergetic processes – the way in which energy is transformed from a source, into energy that can be used by a particular cell, organism, or species. This exchange of energy can occur using energy from the sun, sugars, or from inorganic chemicals. The structure of cell membranes and membrane-bound organelles allow the cells to proceed with optimal efficiency to maintain a dynamic homeostasis within the cell, the organism, and the environment. Let's examine these processes in detail!

Enduring Understanding 2.A. Growth, reproduction and maintenance of the organization of living systems require free energy and matter

2.A.1: All living systems require constant input of free energy

Cellular respiration in eukaryotes involves a series of coordinated enzyme-catalyzed reactions that harvest free energy from simple carbohydrates.

Cellular respiration is the metabolic pathway in which food (e.g., glucose) is broken down to produce energy. Both plants and animals use respiration to create energy required for cellular processes. In respiration, energy is generated during the transfer of electrons in a process known as an **oxidation-reduction (redox)** reaction. The oxidation phase of this reaction is the loss of an electron and the reduction phase is the gain of an electron. Redox reactions are important for all stages of respiration, and the energy that is produced is in the form of adenosine triphosphate (ATP), which we will discuss in detail in section 2.A.2.

Glycolysis is the first stage in cellular respiration. It occurs in the cytoplasm of the cell and does not require oxygen. There are ten stages of glycolysis, each requiring a specific enzyme for catalysis, but you do not need to memorize all of the steps for the exam. The main reaction is illustrated below, and you should make sure to understand and remember the net effects:

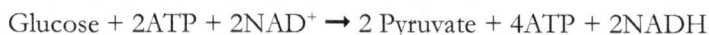

$$\text{Glucose} + 2\text{ATP} + 2\text{NAD}^+ \rightarrow 2 \text{ Pyruvate} + 4\text{ATP} + 2\text{NADH}$$

Glycolysis takes glucose (a six-carbon molecule) and breaks it down into two molecules of **pyruvate,** which contains three carbons each. To start this reaction, two ATP molecules are required, and since the resulting number of ATP molecules produced is four, the net output of ATP molecules by the end of glycolysis is two.

Since glycolysis occurs in the cytosol, the pyruvate enters the mitochondrion by active transport through the mitochondrial membrane where it is oxidized into a compound called **acetyl coenzyme A** (acetyl CoA) through the conversion of one carbon into carbon dioxide during the so-called **Link Reaction**; therefore, acetyl CoA is a two-carbon molecule. Remember that for each molecule of glucose at the beginning of glycolysis, two molecules of acetyl CoA will be produced and utilized in the next stage of respiration (one for each molecule of pyruvate formed in glycolysis).

In prokaryotes, these reactions take place across the plasma membrane. In eukaryotes, the next respiration reactions occur in the mitochondria (mitochondria are discussed in the section on organelles). Now it's time to delve further into the details of mitochondrial structure so that you will understand where ATP is created. Each mitochondrion, as we know, has a double membrane, therefore there are four distinct regions of the mitochondrion to be aware of: the matrix, inner mitochondrial membrane, intermembrane space, and the outer membrane.

The **Krebs cycle** (also known as the citric acid cycle), begins with the generation of a 6-carbon molecule called citric acid, formed from the 2-carbon acetyl CoA and a 4-carbon molecule. Next, two carbons are lost as carbon dioxide (CO_2) and a 4-carbon molecule is formed, which is available to join with acetyl CoA to form citric acid again. Since we started with two molecules of acetyl CoA, two rounds of the Krebs cycle are necessary to process the original molecule of glucose. Finally, eight hydrogen atoms are released and picked up by FAD and NAD (vitamin and electron carriers).

Therefore, for each molecule of acetyl CoA (remember, you started with two) you get

- 1 molecule of ATP
- 3 molecules of NADH
- 1 molecule of $FADH_2$

At this point, the breakdown of cellular respiration results in

- the creation of four molecules of ATP: 2 from glycolysis and 1 from each of the two rounds of the Krebs cycle.
- the release of six molecules of carbon dioxide, two prior to entering the Krebs cycle and two for each of the two turns of the Krebs cycle.
- the generation of 12 carrier molecules (10 NADH and 2 $FADH_2$) that will transport electrons to the next step of aerobic respiration.

Electron Transport

As the Krebs cycle alone does not produce many ATP molecules, its main role is the transfer of electrons that are subsequently used in the electron transport chain to generate large numbers of ATP molecules. The **Electron Transport Chain** uses electrons to pump protons (H^+ ions) across the mitochondrial membrane through a series of proteins including those called **cytochromes**.

The electron carriers we mentioned above, NAD^+ and FAD, have collected their electrons and now these charged carriers, NADH and $FADH_2$, are able to transfer electrons from glycolysis and the Krebs cycle to transform the hydrogen atoms to hydrogen ions and electrons. These high-energy electrons begin their journey through the series of proteins embedded in the inner mitochondrial membrane. At the end of the chain, the electrons encounter the final electron acceptor: oxygen. Combining with the electrons and hydrogen, oxygen forms water. The presence of oxygen (putting the meaning of "aerobic" into "aerobic respiration"), is crucial to the entire process. If there is no oxygen, ATP production would cease, and the cells might not survive.

Chemiosmosis

The electron transport chain does not make ATP directly. Instead, it breaks up a large free energy drop into a more manageable one. While the electrons are passing through the electron transport chain, the hydrogen ions (transformed by NADH and $FADH_2$ at the beginning of oxidative phosphorylation) get pumped into the intermembrane space, generating a proton (H^+) gradient that is used in ATP synthesis. This process is called **chemiosmosis.** During chemiosmosis, the enzyme that creates ATP, the **ATP synthase**, uses the kinetic energy generated by the movement of hydrogen ions through the membrane to phosphorylate ADP. Since the process of these protons (hydrogen ions) diffusing from an area of high concentration to low concentration is similar to water molecules moving across a membrane by osmosis, this process is thus called chemiosmosis.

In total, the oxidative phosphorylation process including the electron transport chain and chemiosmosis produces 34 ATP; the net gain from the whole process of respiration is 36 molecules of ATP.

The electron transport chain captures free energy from electrons in a series of coupled reactions that establish an electrochemical gradient across membranes.

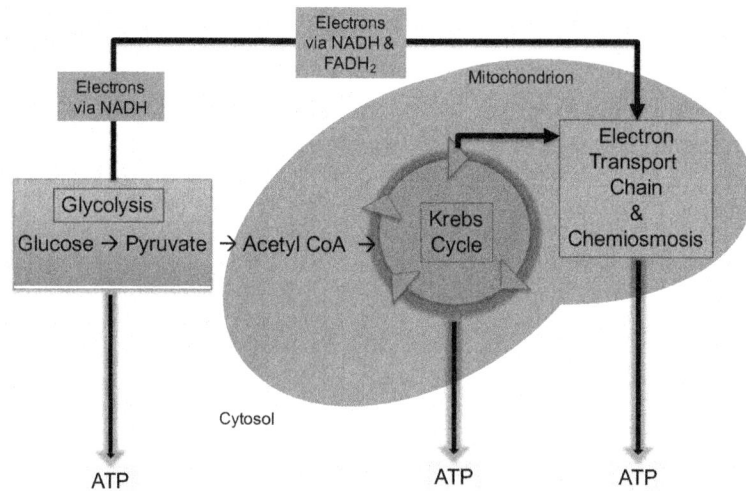

Stages of cellular respiration

Anaerobic respiration – fermentation

Some organisms are anaerobic and cannot use oxygen to make ATP. The production of 2 ATP molecules from the process of glycolysis is a start in their quest to make energy. After glycolysis, anaerobic organisms use **fermentation** to convert pyruvate to lactic acid or alcohol and carbon dioxide.

To produce **alcohol,** yeast and some bacteria convert pyruvate to ethanol (ethyl alcohol) by releasing carbon dioxide from the pyruvate, and reducing acetaldehyde by NADH to NAD^+.

$$\text{Glycolysis} \rightarrow 2\ \text{Pyruvate} + 2NADH \rightarrow 2\ \text{EtOH} + CO_2 + 2NAD^+$$

Other types of bacteria produce **lactic acid**, and they accomplish this by reducing pyruvate by NADH, resulting in lactic acid formation. Lactic acid is also produced in the muscles of animals during exercise to generate more energy. You may have felt the effects of lactic acid after a hard workout—lactic acid buildup in your muscles causes them to be sore!

$$\text{Glycolysis} \rightarrow 2\ \text{Pyruvate} + 2NADH \rightarrow 2\ \text{lactic acid} + 2NAD^+$$

Both aerobic and anaerobic pathways oxidize glucose to pyruvate through the process of glycolysis and both pathways employ NAD+ as an oxidizing agent. A substantial difference between the two pathways is that in fermentation, an organic molecule such as pyruvate or acetaldehyde—rather than oxygen, serves as the final electron acceptor. Another key difference is that aerobic respiration yields much more energy from a sugar molecule than fermentation does. While fermentation is not the most efficient type of respiration (only releasing 2 ATP molecules for each glucose), energy remains stored in lactic acid or alcohol until it is needed.

As we discussed, living systems require free energy to exist. Individual organisms use and store energy in different ways. In this section, we will also discuss how organisms with chloroplasts store and process energy. You will be able to see how the products from photosynthesis are used in the process of cellular respiration described above.

Photosynthesis

Photosynthesis is an anabolic process that stores energy in the form of a six-carbon sugar, for example, glucose. Only organisms that contain chloroplasts (i.e., plants, some bacteria, and some protists) are able to capture free energy from sunlight and use it to produce carbohydrates from carbon dioxide.

An **autotroph** (meaning "self feeder") is an organism that makes its own food—the organic molecules to be used as metabolites from the energy of the sun or other elements. Autotrophs can be divided into two types:

1. **Photosynthetic organisms** make food from light and carbon dioxide, releasing oxygen that can be used for respiration.
2. **Chemosynthetic organisms** obtain energy from small inorganic molecules often in the absence of oxygen.

Heterotrophs (meaning "other feeder") are organisms that hydrolyze carbon compounds produced by other organisms in order to obtain energy, for example, carbohydrates, lipids, and proteins. All animals are heterotrophs.

The **chloroplast** is the site of photosynthesis in a plant cell. Similar to mitochondria in a eukaryotic cell, chloroplasts contain an increased surface area of membrane called the thylakoid membrane. The thylakoid membrane contains pigments (chlorophyll) that are capable of capturing light energy. Between the membranous stacks of thylakoids there is a fluid called **stroma**.

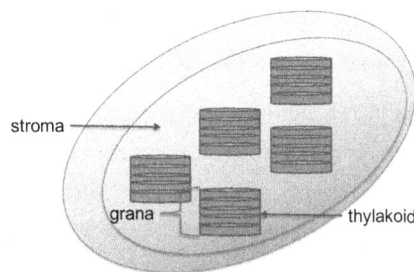

The chloroplast

> Heterotrophs capture free energy present in carbon compounds produced by other organisms; autotrophs capture free energy from physical sources in the environment.

Photosynthetic pigments

The process of photosynthesis begins with the sun. Light is carried as photons, which are fixed quantities of energy. Light is reflected (what we see), transmitted, or absorbed (what the plant uses). Visible light ranges between the wavelengths of 750 nanometers (red light) to 380 nanometers (violet light). As the wavelength decreases, the amount of available energy increases. Plants have several pigments that capture light of specific wavelengths including:

Chlorophyll *a* (the predominant pigment) reflects green and absorbs violet/blue and red light.

Chlorophyll *b* reflects yellow/green light and absorbs blue and red light.

Carotenoid reflects yellow/orange, and absorbs violet/blue light.

The pigments absorb the photons of light as energy, which excites electrons in the chlorophyll. The high-energy electrons are transferred to other molecules called primary electron acceptors, which are located on the thylakoid membrane. These electrons get passed through electron carriers, resulting in the production of ATP (energy) and NADPH (electron carrier).

Photosynthesis reverses the electron flow we saw in the previous chapter on cellular respiration. Water is split by the chloroplast into hydrogen and oxygen. Oxygen is released as a waste product as carbon dioxide is reduced to sugar $C_6H_{12}O_6$ (glucose). This requires the input of energy, which comes from the light.

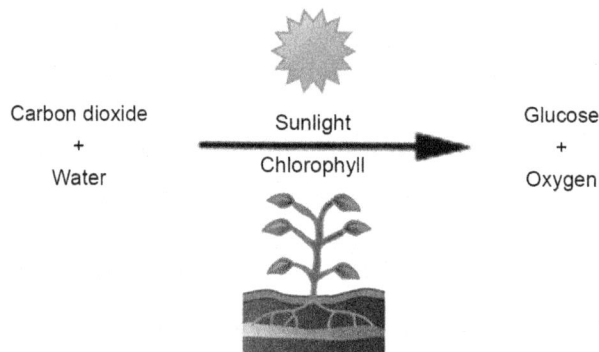

Carbon dioxide	Sunlight	Glucose
+	Chlorophyll	+
Water		Oxygen

Photosynthesis scheme

The formula for photosynthesis is:

$$CO_2 + H_2O + energy\ (from\ sunlight) \rightarrow C_6H_{12}O_6 (glucose) + O_2$$

Photosystems

The combination of electron acceptors and the pigments forms photosystems bound to the thylakoid membrane that are capable of capturing light energy. **Photosystems** are units of protein complexes that contain a **reaction center** that releases an electron to the primary **electron acceptor** after stimulation by the pigment. This transfer is the first step of the light reactions. There are two photosystems, named accordingly by their date of discovery, not their order of occurrence, therefore, in the sequence of photosynthesis reactions, Photosystem II occurs before Photosystem I!

Photosystem I is composed of a pair of chlorophyll *a* molecules. Photosystem I is also called P700 because it absorbs light at a wavelength of 700 nanometers. Photosystem I makes ATP whose energy is needed to synthesize glucose.

Photosystem II is also called P680 because it absorbs light at a wavelength of 680 nanometers. Photosystem II produces ATP + $NADPH_2$ and oxygen gas as a by-product.

Light-dependent reactions—the light reactions

Photosynthesis occurs in two stages, the light reactions and the Calvin cycle (dark reactions). The conversion of solar energy to chemical energy occurs during light reactions. In light reactions, chlorophyll can absorb light and use the energy to split water, releasing oxygen as a by-product. The conversion of light energy to chemical energy is stored in the form of NADPH and ATP. Both NADPH and ATP are then used in the Calvin cycle to produce sugar.

The light-dependent reactions of photosynthesis in eukaryotes involve a series of coordinated reaction pathways that capture free energy present in light to yield ATP and NADPH, which power the production of organic molecules.

The production of ATP is termed **photophosphorylation** due to the use of light, ADP, and phosphates. Depending on the pattern of the electron transport, two types of photophosphorylation, exist: cyclic, and non-cyclic.

Cyclic Photophosphorylation

Cyclic photophosphorylation occurs predominantly in bacteria, or in plants when there are not enough NADP molecules to accept the electrons. Therefore, during cyclic photophosphorylation, only ATP is produced. In Photosystem I, upon stimulation by light, the electrons are raised to an excited level, then transferred through carrier proteins of the electron transport chain, until they eventually return back to the P700 photosystem. Photosystem I is therefore appropriately called cyclic photophosphorylation because the electron pathway occurs in a cycle as shown below. Photosystem I is also a component of non-cyclic photophosphorylation, which starts with light and ends with glucose.

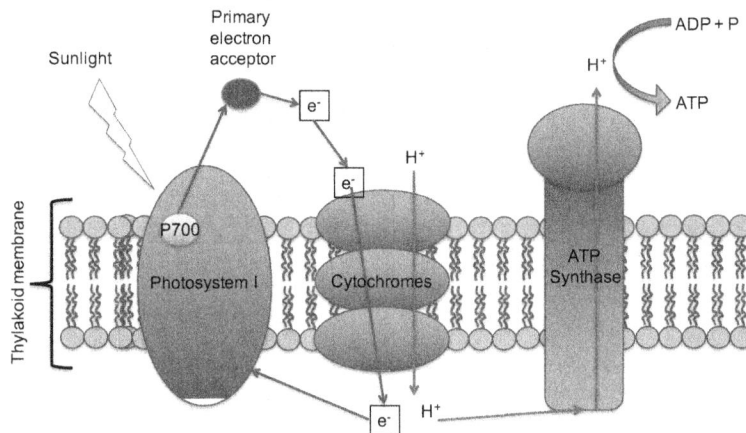

Cyclic Photophosphorylation

Non-cyclic Photophosphorylation

Non-cyclic photophosphorylation predominantly occurs in green plants. It produces ATP and NADPH using both Photosystem I and Photosystem II. The process starts when sunlight captured by a leaf excites the reaction center of Photosystem II, P680. Simultaneously, the light absorbed by P680 also results in **photolysis**—the separation of water molecules into hydrogen ions, oxygen, and electrons. The activated electrons travel to the primary acceptor and through the electron transport chain through proteins called cytochromes. Hydrogen ions move through ATP synthase to produce ATP. When the electrons arrive at Photosystem I, they become activated and passed through the electron transport chain combining with $NADP^+$, resulting in the generation of NADPH.

Now that you have learned about both non-cyclic and cyclic photophosphorylation, you will notice that Photosystem II only participates in non-cyclic photophosphorylation.

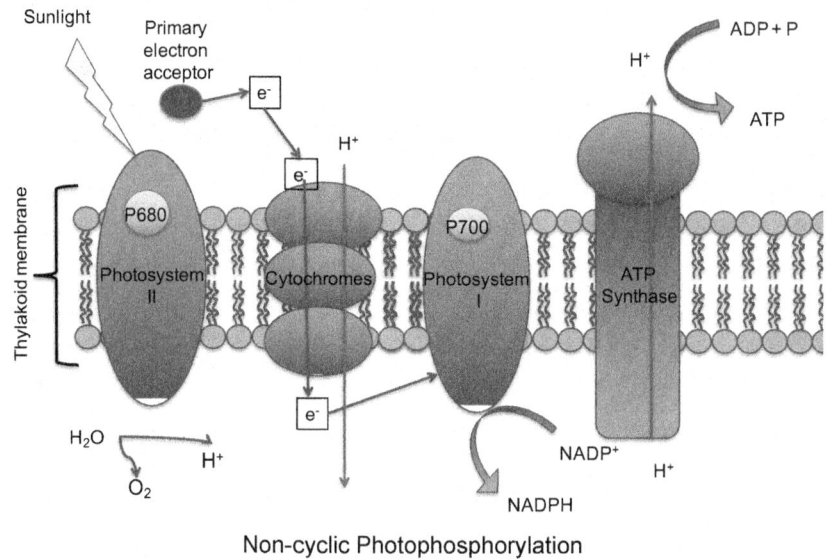

Non-cyclic Photophosphorylation

The chart below outlines the differences between cyclic and non-cyclic photophosphorylation.

Cyclic photophosphorylation	Non-cyclic photophosphorylation
Uses only Photosystem I	Uses Photosystems I and II
e⁻ travel back to Photosystem I	e⁻ from Photosystem I accepted by NADP⁺
No photolysis	Photolysis produces O_2 as a by-product
Reactions occur in thylakoids	Reactions occur in thylakoids

The Calvin cycle

The next stage of photosynthesis includes the process by which glucose is formed, called the **Calvin cycle** and occurs in the stroma of the chloroplast. **Carbon fixation** facilitates the formation of carbohydrates that will ultimately form sugars (typically glucose). Carbon dioxide from the air is incorporated into organic molecules present in the chloroplast. The first carbon compound formed from carbon dioxide contains three carbon atoms. Most plants use a single enzyme, **ribulose bisphosphate carboxylase** (rubisco), to collect carbon dioxide from the air and carry out photosynthesis. ATP and NADPH from the light reaction are required to convert carbon dioxide to carbohydrate (sugar).

A summary figure of the processes that work together to transform energy from the sun into chemical energy is on the next page:

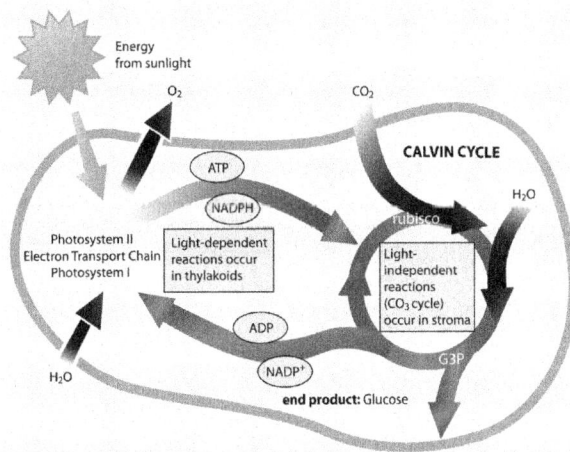

Energy of the sun is transformed into chemical energy in plant cells

To further put all of these processes into perspective, we have provided a diagram of the relationship between photosynthesis and cellular respiration.

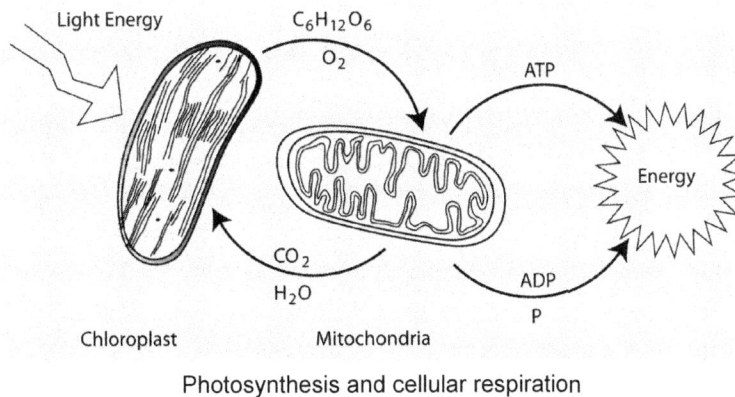

Photosynthesis and cellular respiration

Photorespiration

Photorespiration is a light-dependent process and is the reverse of photosynthesis: oxygen is taken up from organic compounds, and carbon dioxide is released. During photorespiration, carbon fixation becomes less efficient because the concentration of carbon dioxide is reduced. This results in the fixation of oxygen.

Modification for dry environment

Plants that are adapted for hot and arid climates produce energy using a strategy to reduce water loss. The carbon fixation **crassulacean acid metabolism** or **CAM photosynthesis** is a specialized form of energy production. At night, the stomata in the leaves of the plant are opened and CO_2 is stored as malate in the vacuoles. During the day, the stomata are closed to reduce evapotranspiration while the stored CO_2 enters the Calvin cycle.

Thermoregulation

Heat is another form of energy, the thermal energy. The way animals regulate their thermal energy is an important component of their biology. There are two main types of

Photosynthesis first evolved in prokaryotic organisms; scientific evidence supports that prokaryotic (bacterial) photosynthesis was responsible for the production of an oxygenated atmosphere; prokaryotic photosynthetic pathways were the foundation of eukaryotic photosynthesis.

thermoregulation: ectothermy and endothermy. **Ectothermy** refers to when an animal relies on the external environment for temperature control instead of generating its own body heat through metabolic processes. This form of thermoregulation is sometimes called "cold-blooded" though really the temperature of the animal, and its blood, are in a linear relationship. The majority of all animal life are ectotherms, including invertebrates, amphibians, and nearly all fish and reptiles. Because ectotherms depend on the temperature of their environment, they cannot move much unless the ambient temperature is optimal. **Endotherms** on the other hand, self-regulate their temperature by producing their own heat through metabolism. Birds, mammals, marsupials, and some active fish like sharks and swordfish all self regulate their body temperature at a near constant temperature regardless of their ambient temperature. Shivering, sweating, bathing in water or basking in the sun are all examples of behavioral adaptations that animals employ to help regulate their body temperatures.

The food chain

All organisms need to obtain their energy for life from another source. Plants get energy from the sun, animals eat plants, and some animals eat other animals. This sequence of who eats whom in a biological community is called a **food chain**. It always starts with the primary producers or autotrophs. They are using the simplest form of abiotic energy to grow. Plants, algae and phytoplankton,—all depend on the sun for photosynthesis. In deep sea hydrothermal vents where there is no sunlight, chemosynthetic bacteria make their food from the chemicals spewing from the Earth's crust. Next, the primary consumers eat the autotrophs. This would be like a mouse eating some grass. Then a secondary consumer, like a snake, would eat the mouse. Then a tertiary consumer, like an owl, would eat the snake. The food chain continues until the top predator is reached which is an animal that has no natural enemies as a full grown individual. The energy flows from the abiotic source all the way to the top of the food chain.

Changes in sunlight variations, through seasonality or geography, can impact the abundance of the primary producers. This variation impacts the rest of the chain above the autotrophs. In temperate climates for example, in the fall, plant life has less sun available. Less green leaves means less food for insects and other herbivores, resulting in less food available for the higher predators. Birds migrate south, where warmer conditions and rainy seasons provide an abundance of food. Other vertebrates hibernate until there are more resources available. In the spring, new plant growth provides new resources and expands the food chain to a higher capacity. This encourages reproduction of animals and plants due to the greater energy availability.

2.A.2: Organisms capture and store free energy for use in biological processes

ATP-adenosine triphosphate

Maintaining life through growth and reproduction of living systems requires energy. The **first law of thermodynamics** states that energy can be transferred or transformed, but it cannot be created or destroyed. Therefore, we need to know from where the energy comes from to maintain the order in living systems. In contrast, **entropy** is disorder in these systems. The **second law of thermodynamics** states that entropy increases over

Life requires a highly ordered system, but living systems do not violate the second law of thermodynamics, which states that entropy increases over time.

time, therefore, life has evolved cellular processes that balance between the fluctuations in entropy, thereby offsetting negative changes in free energy with positive changes in free energy. Thus, **exergonic** (energy-releasing) reactions counteract **endergonic** (energy-absorbing) reactions to maintain an equilibrium. We will explain these processes in detail below, but first, we need to discuss the molecule that is responsible for intracellular energy transport: adenosine triphosphate, or **ATP!**

ATP contains an adenosine molecule bound to three phosphate groups. The phosphate bonds store a large amount of energy. This is very convenient for a cell needing energy: breaking a phosphate bond results in the release of energy and the formation of adenosine *di*phosphate (ADP) and a phosphate group. An exergonic reaction such as this is used to maintain order in a living system:

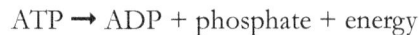

$$ATP \rightarrow ADP + phosphate + energy$$

As we have already discussed earlier in section 2.A.1, there are two types of cellular respiration: aerobic, when ATP is produced in the presence of oxygen; and anaerobic, when ATP is produced in the absence of oxygen.

2.A.3: Organisms must exchange matter with the environment to grow, reproduce and maintain organization

Water is a major player in many chemical reactions. We briefly discussed that water is a polar molecule because it has a negatively charged end and a positively charged end. This property allows water molecules to dissolve many kinds of chemicals. There are four distinct properties of water that contribute to its importance in facilitating living systems function:

- **Cohesion** — is the principle by which hydrogen bonds can hold liquid particles together. Think about a dew drop, which is made up of many water molecules that keep themselves together by cohesive forces. Cohesion is responsible for the phenomenon called surface tension, which gives the water its unique properties.
- **Adhesion** – is the ability of a substance to stick to a different type of substance.
- **Heat capacity** – is the ability of a substance to store heat, which also asks how much heat is required to change the temperature by one degree. Water has a very high heat capacity, and therefore is important to life because it can help maintain body temperature.
- **Expansion on freezing** – The ability of water to expand when it freezes, also makes it stand out as an important molecule. Most molecules get closer together when they cool from liquid to solid, but this is the opposite in water. This is important because it means that ice floats on small bodies of water, which allows life to survive under the ice!

Surface area-to-volume ratios affect a biological system's ability to obtain necessary resources or eliminate waste products

Enduring understanding 2.B: Growth, reproduction and dynamic homeostasis require that cell create and maintain internal environments that are different from their external environments

2.B.1: Cell membranes are selectively permeable due to their structure.

Prokaryotes have a thick **cell wall** (made up of cellulose, other polysaccharides, and proteins) that protects the cell, provides shape, and prevents the cell from bursting. The cell wall is thick enough for support and protection, yet porous enough to allow water and dissolved substances to enter the cell. The antibiotic medication penicillin targets and disrupts cell wall synthesis, thereby killing the cell. Some prokaryotes have a capsule, (made of polysaccharides), surrounding the cell wall for extra protection from more complex organisms.

Cell membranes separate the internal environment of the cell from the external environment and allow molecules to pass in and out of the cell. The presence of phospholipids allows the membrane to be both hydrophilic and hydrophobic. As pictured below, the phosphate heads are oriented toward the aqueous environments found externally and internally, and the fatty acid tails are oriented toward each other on the interior of the membrane itself.

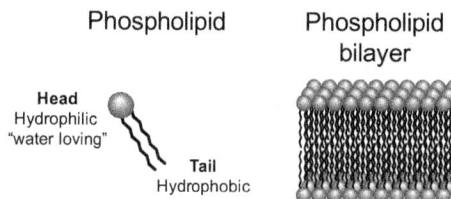

Phospholipid   Phospholipid bilayer

Head
Hydrophilic
"water loving"

Tail
Hydrophobic

This phospholipid membrane also contains proteins embedded to varying degrees—peripheral, integral, and transmembrane proteins that can be **hydrophilic**, with charged and polar side groups, or **hydrophobic**, with nonpolar side groups. The **fluid mosaic model** refers to the flexibility of the phosophlipid membrane to contain these proteins that allow molecules to enter and leave the cell. Carbohydrate side chains and cholesterol molecules also help the cell membrane maintain its stability.

These basic molecules: carbohydrates, lipids, and proteins, will be discussed in detail in Chapter 4.

2.B.2: Growth and dynamic homeostasis are maintained by the constant movement of molecules across membranes.

Since the cell membrane is selective in terms of what is allowed to cross, the cells are able to have internal environments that are different from their external environment. This **selective permeability** maintains dynamic homeostasis by various modes of transport including passive and active transport. Hydrophilic substances such as large polar molecules and ions have the ability to get in and out of the cell by moving across the membrane through embedded channels and transport proteins. Even though water is polar, it quickly moves across the lipid bilayer through channel proteins called **aquaporins**.

Selective permeability is a direct consequence of membrane structure, as described by the fluid mosaic model.

Metabolic energy is not required for **passive transport** because it entails spontaneous movement of molecules down the concentration gradient. **Simple diffusion** allows particles to move from high to low concentrations, and **osmosis** allows movement of water down the concentration gradient of the solute, for example a sugar.

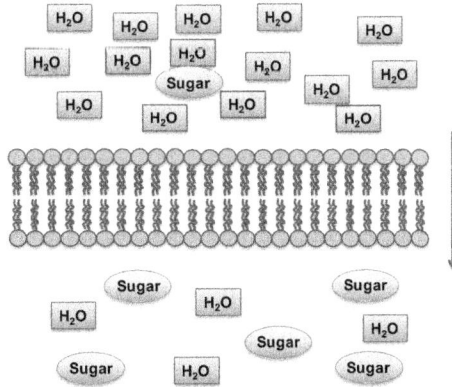

During osmosis, the water molecules move from low sugar concentration to the high sugar concentration.

Specialized proteins in the membrane, called **channel proteins**, help lipid-insoluble (charged and polar) molecules to move down the concentration gradient in a passive transport process called **facilitated diffusion**. Another type of integral membrane protein called **carrier proteins** bind a molecule on one side of the membrane, undergo a conformational change, and release the molecule to the other side of the membrane in a process called active transport.

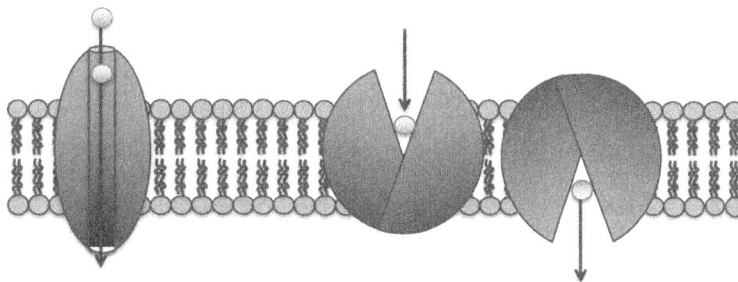

Channel Protein   Carrier Protein

Types of transport

Active transport requires metabolic energy and transport proteins to move molecules from low to high concentrations across membrane. Active transport establishes concentration gradients vital for dynamic homeostasis, including sodium/potassium pumps in nerve impulse conduction and proton gradients in electron transport chains during photosynthesis and cellular respiration. The process of **endocytosis** facilitates movement of large molecules from the external environment to the internal environment, in which the membrane can bud inward to engulf outside material. In contrast, **exocytosis** moves large molecules from the internal environment to the external environment. Exocytosis is a secretory mechanism, (the reverse of endocytosis), and the way, for example, to get protein produced in the cell out to the circulation.

Active transport requires free energy to move molecules from regions of low concentration to regions of high concentration.

2.B.3: Eukaryotic cells maintain internal membranes that partition the cell into specialized regions.

The **cell membrane** encloses the cytoplasm and consists of a lipid bilayer that controls the passage of molecules in and out of the cell. Outside of the cell membrane, prokaryotes have a thick cell wall but generally lack internal membranes. The cell has internal membranes that separates organelles, and other subcellular structures. We will discuss the functions of the organelles later in this book. Here we will explore the role of internal membranes in isolating chemical reactions that would otherwise compete with each other. First, as mentioned above, it is important to note that prokaryotes do not have membrane-bound organelles, but they do contain vesicles that help determine structure and function of the cell, as well as serving to contain enzymes until they are needed. Eukaryotes on the other hand, do contain organelles and a complex membrane system. This allows the cells to function together and grow into larger organisms. Furthermore, whereas prokaryotes do not contain nuclei, eukaryotes contain nuclei that are surrounded by a nuclear membrane to prevent the free passage of molecules between the nucleus and the cytoplasm. Mitochondria also contain an inner membrane that separates the mitochondrial matrix from the intermembrane space. This inner mitochondrial membrane contains the complexes of the electron transport chain, ATP synthase complex, and transport proteins, so it is therefore very important for energy production. **Vacuoles** are also membrane bound and can be found in plant cells to hold stored food and pigments. The large size of vacuoles allows them to fill with water in order to provide turgor pressure, without which a plant will wilt.

Enduring understanding 2.C: Organisms use feedback mechanisms to regulate growth and reproduction, and to maintain dynamic homeostasis

2.C.1: Organisms use feedback mechanisms to maintain their internal environments and respond to external environmental changes.

Organisms can control various functions through feedback mechanisms in which products of a reaction are able to control the activity and regulate the further production of a substance. Such feedback can be positive or negative.

Control of gene expression, or transcription, has been largely discovered by studies in bacterial cells. An **operon** is a portion of DNA that acts as a regulatory system, able to **induce** or **repress** gene systems. Small molecules interacting with an operon can inhibit gene expression when a repressor binds to the DNA sequence, or stimulate gene expression when an inducer binds to the DNA sequence.

An example of a functional operon is the *lac* operon, which contains the genes coding for enzymes used to catabolize lactose. Three different genes involved in digestion of lactose are coded for within the lac operon: *lac Z*, *lac Y*, and *lac A*. The *lac* operon contains a promoter and an operator that is the "off and on" switch for the operon. In the case of the *lac* operon, when lactose is absent, the repressor is active and the operon is turned off. The operon is turned on again when allolactose (a molecule formed from lactose) inactivates the repressor by binding to it.

In contrast, some genes are always activated! For example, the *trp* operon encodes for the amino acid tryptophan. As illustrated in the figure below, tryptophan will bind to the repressor when levels are high, preventing the synthesis of more tryptophan. This is an example of negative feedback that allows an organism to keep it physiologically in balance.

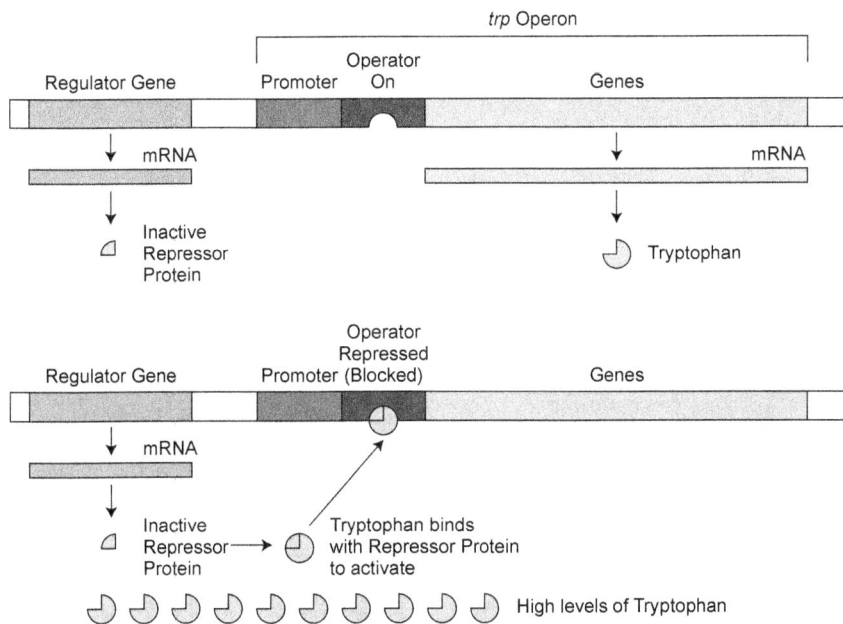

Another example of a feedback mechanism is when water is not readily available for plants. In most plants, the first step in the light-independent reactions of photosynthesis involves the fixation of carbon dioxide by rubisco to eventually produce a three-carbon sugar in C3 photosynthesis, which takes place throughout the leaf. To enable a more efficient use of water in hot, arid environments with intense sunlight, some plants employ C4 photosynthesis, which produces a compound containing four carbon atoms. C4 photosynthesis is an adaptation that helps prevent photorespiration and uses another enzyme, **phosphoenolpyruvate carboxylase** (PEP carboxylase) to facilitate carbon dioxide delivery to the rubisco enzyme.

The process starts when carbon dioxide enters the plant at the mesophyll cells, the internal leaf cells and joins with another carbon compound to make a four-carbon acid. This four-carbon chain is easily broken into a three-carbon compound and carbon dioxide that can be transferred to the chloroplast to undergo processing in the Calvin cycle to make glucose. Faster and more efficient uptake and delivery of carbon dioxide allows the plant to keep its stomata closed for a longer time, limiting water loss, while still acquiring adequate amounts of carbon dioxide. In this way, these plants gain a competitive advantage over other plants because they can promote carbon fixation under conditions of extreme temperatures, drought, or limitation of nitrogen and carbon dioxide.

Positive feedback, on the other hand, allows for an increase in the formation of a product. Prolactin is a hormone which activates milk production in lactating mammals. The process of suckling by the infant mammal stimulates the production of more prolactin, which further increases milk production. Therefore, this is an example of a positive feedback mechanism.

Negative feedback mechanisms maintain dynamic homeostasis, returning the changing condition back to its target set point, while positive feedback mechanisms amplify responses and processes in biological organisms.

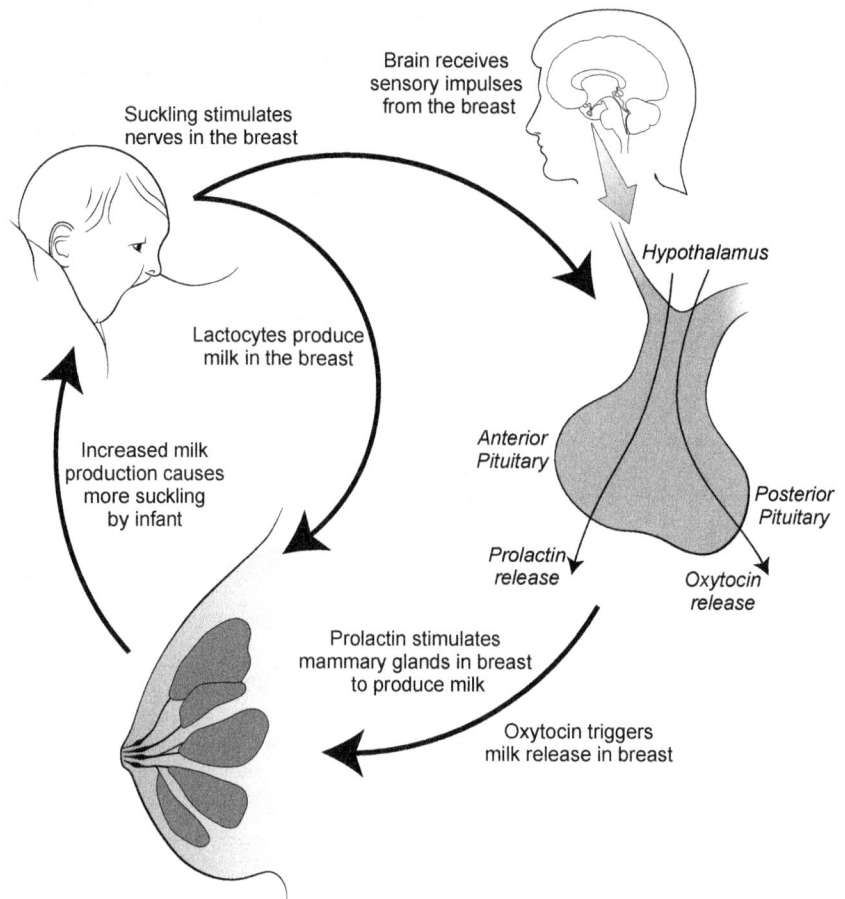

Abnormalities in a feedback mechanism can result in disease. Let's take the thyroid gland as an example, whose role it is to regulate metabolism in tissue by a hormone called thyroxine. Decreased production and release of thyroxine can lead to hypothyroidism, which is characterized by a slow metabolic rate, resulting in weight gain and exhaustion since energy is not produced sufficiently.

2.C.2 Organisms respond to changes in their external environments.

There are also other ways in which the organisms use their behavior and physiology in order to respond to changes in their environments.

Beginning with plants, a **photoperiod** is the duration of a plant's daily exposure to light. Variations in the length of photoperiods affect its growth, development, and physiological processes. For example, plants adapt to seasonal changes in photoperiods by increasing or decreasing growth processes, and changing patterns of photosynthesis and respiration. In fact, there are some plants that are so specialized in their photoperiods, that they cannot survive unless they receive a certain amount of light. Some species of plants develop traits over time, allowing them to thrive in the characteristic photoperiods of their native environment. **Phototropism** is when a plant grows toward the light – it is responding to a stimulus. A classic experiment demonstrating this principle is performed by placing a bright light on one side of a plant. Within a short time, the plant will bend toward the light. When the plant is turned around, facing away from the light , it will bend back toward the light within a short time!

Movement in a cell or an organism is called **taxis**, if the response is a directional movement to a stimulus. For example, the detection of pheromones by a male animal will prompt taxis towards the female. Another form of movement, **kinesis**, is not directional but instead, a random movement. As an example, think of a paramecium moving around its environment without any direction. Therefore, more complex behavior is exhibited with taxis compared to kinesis.

Since behaviors are usually carried out as the result of a stimulus, the process of hibernation is a prime example - when the temperature drops and the days get shorter, bears know that it is time to start their hibernation period. In a similar way, cycles of light and dark also regulate activities. **Diurnal** animals are active during the day, and **nocturnal** animals are active at night. These daily cycles of activity are built-in to the physiology of plants and animals, and are known as the **circadian rhythm**, which is adjusted according to external cues.

Temperature changes also elicit different physiological responses in humans and other mammals. Shivering in response to cold is a contraction of the muscles in order to increase the metabolic rate and generate heat. On the other hand, sweating in response to heat decreases your body temperature by the process of evaporating sweat from your skin – providing a cooling effect.

Enduring understanding 2.D: Growth and dynamic homeostasis of a biological system are influenced by changes in the system's environment

2.D.1: All biological systems from cells and organisms to populations, communities and ecosystems are affected by complex biotic and abiotic interactions involving exchange of matter and free energy.

As we just discussed, organisms evolved to have machinery dedicated to regulation for internal and external changes. As an additional factor, organisms as a biological system or a group have mechanisms by which they are able to regulate changes in response to environmental factors. In each ecosystem, there are **biotic** (living) and **abiotic** (nonliving) factors. At the cellular level, abiotic factors such as the amount of sunlight, the temperature, and the availability of gases or water, dictate the extent to what type of organisms exist, and their ability to reproduce. For example, the extreme cold of the Arctic is not conducive to the growth of trees or plants because of the frozen state of the soil.

Cell activities, organism activities, and the stability of populations, communities and ecosystems are affected by interactions with biotic and abiotic factors.

Populations, communities and ecosystems

Complex biotic and abiotic interactions involving exchange of matter and free energy are the underlying principles that lead natural selection to give us the diversity of life seen across Earth's history. We previously discussed the importance of the food chain in ecosystems and community interactions. As energy moves up the food chain, the form of the matter changes. Autotrophs convert the free energy of the sun or hydrothermal vents into usable building blocks for life. These are consumed and transformed again by the consumers on each level of the food chain. The availability of energy and matter in an ecosystem is what sets the levels of biodiversity that can be supported. Changes in the availability of that energy or novel niches can change this capacity in either way.

- If more energy is available in the food chain, then more diversity is supported
- If less energy is available, the levels of diversity supported are less.

This gradient of overall biodiversity is seen in ecosystems across the globe. At the equator, where solar energy is at its peak throughout the year, you see greater diversity supported. As you move away from the equator toward the poles you see a decrease in biodiversity.

Competition between individuals is where successful energy harvesting plays a key role in the success of the individual. For example, cave swallows in China are able to build their nests on ledges of the mouth of the cave. They use their saliva like glue as they work gathered materials into their nest. The walls of the cave have limited real estate for building a stable nest that will rear young. Because mates are found in that cave, it is ideal to invest energy into building a successful nest. So, only mating pairs that can invest in establishing a good nest, which means having enough energy to do so, will be successful in producing young. This competition can have an impact on the genetic diversity of a population.

- **Intense competition/limited resources (low energy availability)**
  - ⟶ only individuals with best genes are likely to be successful
- **Less competition/abundant resources (high energy availability)**
  - ⟶ more genetic diversity is observed in the population

2.D.2: Homeostatic mechanisms reflect both common ancestry and divergence due to adaptation in different environments.

Homeostasis is the process of balancing a particular condition in the internal environment at equilibrium, regardless of external changes. It is important to the regulation of physiological functions in organisms. However, to achieve homeostasis, organisms have adapted to their particular home environments. Each is challenged to maintain metabolic rates, nutrient absorption and waste excretion, and gas exchange at nominal levels to ensure fitness within its environment. For example, aquatic and terrestrial plants and animals must exchange CO_2 and O_2 for proper metabolic respiration. Organisms that live on land have high availability of the gaseous forms of these chemicals, however they run the risk of water loss through respiration. Organisms in aquatic environments have a reduced amount of dissolved gases in their environment but plenty of water. Each have adapted to these conditions in their own way. Aquatic animals use exposed structures, such as gills, that have their oxygen-carrying body fluid in near direct contact with the dissolved gas in the water. This allows for an exchange of carbon dioxide and oxygen. Terrestrial animals would risk dehydration in such a system, so their respiratory tracts are enclosed to reduce water loss. However, the principles are similar. In aquatic plants, because the diffusion of dissolved carbon dioxide is low in the water, their structures tend to be simplified to allow for optimal diffusion of the gasses into the plant tissues. Terrestrial plants have to limit their gas exchange to stomata-regulated pores to reduce water loss. Because of the

Continuity of homeostatic mechanisms reflects common ancestry, while changes may occur in response to different environmental conditions.

mechanisms for utilizing oxygen and carbon dioxide in metabolism, we can infer common ancestors in both terrestrial and aquatic animals and plants, respectively.

In animals, though there are different feeding strategies (carnivore, herbivore, and omnivore) there is a commonality in the mechanisms for digesting food and excretion of nitrogenous waste. All digestion in animals occurs internal to the body cavity and involves digestive enzymes that break down the food into simpler components to be absorbed, metabolized, or excreted. The proteins, amino acids, nucleic acids, and nitrogenous bases are all broken down into their amino groups for absorption. Then, depending on the organism, the nitrogenous waste is excreted into the environment. In most aquatic animals, including many fish, ammonia is the final waste product that is continuously released in the surrounding water. In mammals, amphibians, sharks and some fishes, urea is produced and released in concentrated excretions. For birds, insects, reptiles and land gastropods, uric acid is released, which tends to be the most concentrated form of nitrogenous waste.

Finally, thermoregulation in animals is necessary for both endotherms and ectotherms to maintain homeostasis. We previously discussed the mechanisms used for both to regulate body temperature, but one feature that both groups use is countercurrent heat exchange. This is a way to increase the thermal efficiency of an organism. Imagine a goose with its feet in the water. There are no feathers on its feet and little muscle. The ambient water is cooler than the body temperature of the bird but it is efficiently retaining the heat it generates because as the warm blood leaving the body of the goose travels to the feet, it exchanges its heat with the cooled blood coming from the feet. In this way, body heat is not lost as quickly to the water and decreases the metabolic requirements on the goose while it is swimming.

2.D.3: Biological systems are affected by disruptions to their dynamic homeostasis.

As we have discussed, adaptations regulating homeostasis in organisms are important factors contributing to their fitness within their environment. For example, animals in arid ecosystems have specialized anatomy and physiology to reduce dehydration. Camels, for example, use fat stores to produce water and the excessive heat generated during this process is released at night when the ambient temperatures are cooler. However, an organism lacking these arid environment adaptations will dehydrate or overheat.

Other physiological systems that disrupt homeostasis include responses to pathogens, toxins, and allergens. Toxins can have an effect on the physiology and function of cellular pathways that are essential to homeostasis. Neurotoxins, for example, can block calcium ion channels, disrupting the action potential of the cell, and neutralizing its ability to send a signal. The immune response to pathogens or allergens is also disruptive to homeostasis. Often the release of inflammation-associated proteins and cells, such as cytokines and macrophages, are the cause of swelling and pain. These lead to elimination of the antigen and healing but are not a homeostatic state for the organism.

Homeostasis in ecosystems

Homeostasis in ecosystems is similar to the equilibrium seen in organisms. If there is a balance of the energy within the system, then the food chain is balanced as well. If a component of the ecosystem starts favoring one group over another, shifts would

Disruptions at the molecular and cellular levels affect the health of the organism; disruptions to ecosystems impact the dynamic homeostasis or balance of the ecosystem.

occur in the populations that make up the ecosystem. These shifts can occur in short periods of time, but long-term changes in the ecosystem can lead to permanent changes. For example, when natural disasters or humans impact an ecosystem, they can disrupt the balance therein. Say a logging company reduces the canopy density in a forest, then more sunlight can reach the forest floor. This changes the dynamics in unforeseen ways. Lower vegetation will compete differently, and temperature and humidity conditions can change, which can result in lasting effects on the ecosystem. Likewise, forest fires may sweep through the lower underbrush of a forest and leave only the older trees, leaving a permanent change on the ecosystem. Human activity can also introduce new invasive organisms to an ecosystem that will disrupt the balance as it competes with local flora and fauna to which it was not previously adapted.

2.D.4: Plants and animals have a variety of chemical defenses against infections that affect dynamic homeostasis.

One of the mechanisms of maintaining homeostasis in animals and plants includes the ability to defend against infections. Throughout the animal kingdom, organisms have evolved ways of defending themselves against invading parasites and pathogens. Some of the chief methods are phagocytosis (engulfing and destroying) the invader, encapsulating it, or poisoning it with chemicals.

The immune system in plants

In plants, defensive strategies depend on innate immunity for pathogens as well as toxins to prevent parasitism and herbation. This latter strategy is common, though energetically costly to produce. In cinnamon trees for example, the potent cinnamon compounds make the wood unfavorable to insects. In a similar way, the termites in conifer trees are toxic to many animals at low doses. Hemlock is another poisonous plant that has historically been used to poison the foods of royalty, though in nature, the plant only produces the poison to deter predators. These toxins however, serve little function other then defense, and should not be toxic to the plant itself in order for the energetic investment in producing the toxin to be worth it.

The human immune system

The human immune system has two types of defense mechanisms against foreign invaders: non-specific and specific.

The **non-specific** immune mechanism is composed of two parts. The body's physical barriers—the skin and mucous membranes—are the first line of defense. As long as there are no abrasions on the skin, it prevents the penetration of bacteria and viruses. Mucous membranes form a protective barrier around the digestive, respiratory, and genitourinary tracts. In addition, the pH of the skin and mucous membranes inhibit the growth of many microbes. Mucous secretions (tears and saliva) wash away many microbes and contain lysozymes that kill many microbes.

The second component of the non-specific immune response includes white blood cells and inflammatory responses. Some white blood cells (neutrophils, eosinophils, and monocytes) engage in **phagocytosis**, the process of using the cell membrane to engulf

Plants, invertebrates and vertebrates have multiple, non-specific immune responses.

foreign particles and form an internal (neutralized) phagosome. Neutrophils make up about 70% of all white blood cells. Monocytes mature to become macrophages, which are the largest phagocytic cells.

Instead of killing the invading microbe directly, natural killer cells destroy the body's own infected cells. During an inflammatory response, blood supply to the injured area is increased, causing redness and heat. Swelling typically also occurs with inflammation. Histamine is released by basophils and mast cells when cells are injured, triggering the inflammatory response.

The **specific** immune mechanism recognizes specific foreign material (individual pathogens) and responds by destroying the invader. An **antigen** is any foreign particle that elicits an immune response. The body manufactures an **antibody** to recognize and latch onto antigens and destroy them. Memory of the invaders provides immunity upon further exposure.

Immunity is the body's ability to recognize and destroy an antigen before it causes harm. Active immunity develops after recovery from an infectious disease (e.g., chickenpox) or after a vaccination (e.g., mumps, measles, and rubella). Passive immunity may be passed from one individual to another and is not permanent. A good example is the immunity passed from mother to her nursing child. A baby's immune system is not well developed and the baby is provided additional protection from the the passive immunity he or she receives through nursing.

The immune system attacks not only microbes, but also cells that are not native to the host such as skin grafts, organ transplantations, and blood transfusions. Antibodies to foreign blood and tissue types already exist in the body. If blood is transfused that is not compatible with the host, these antibodies destroy the new blood cells. There is a similar reaction when tissue and organs are transplanted.

The **major histocompatibility complex** (MHC) is responsible for the rejection of tissue and organ transplants. This complex is unique to each person. Since cytotoxic T cells recognize the MHC on transplanted tissue or organ as foreign and destroy these tissues, a transplant recipient needs various drugs to suppress the immune system and prevent rejection of foreign tissue; however, this also leaves the patient more susceptible to infection.

Diseases can occur when the body's immune system goes awry. Instead of defending the body against foreign antigens, it sees the healthy cells as foreign and attacks them! Examples of autoimmune diseases include celiac disease, in which there is a reaction to gluten; Grave's disease, in which there is thyroid dysfunction; vitiligo and psoriasis, which are skin conditions; inflammatory bowel disease; and Type 1 diabetes, in which the cells of the pancreas that produce insulin are destroyed. There is no way to prevent autoimmune disease, but there is extensive research being performed to determine the exact causes and novel treatments for many autoimmune diseases. The research directions for treatments includes inhibiting specific pro-inflammatory proteins in order to decrease inflammation in rheumatoid arthritis for example; as well as research to potentially regenerate pancreatic cells responsible for insulin production, in order to treat diabetes.

Enduring understanding 2.E: Many biological processes involved in growth, reproduction and dynamic homeostasis include temporal regulation and coordination _____

2.E.1: Timing and coordination of specific events are necessary for the normal development of an organism, and these events are regulated by a variety of mechanisms.

Morphogenesis is the name given to the process by which bodily form is established (*morphos* means 'form' and *genesis* means 'origin'). One of the mechanisms to shape the emerging body parts involves programmed cell death, also termed **apoptosis**. Apoptosis occurs widely during early development, to facilitate shape formation of a body part such as fingers and toes, which begin fused as a single projection of cells. As morphogenesis continues, apoptosis occurs between the digits as they elongate. Eventually the digits no longer fused. The order of this morphogenetic process as well as the entire developmental process is due to timed gene expression. This refers to activation and generation of gene products during particular periods of development. The product could be an enzyme, a structural protein, or a control molecule, but their concentrations and locations within the developing embryo determine its appropriate body plan. One example includes the HOX genes. These genes are known as **homeotic genes** that control the body plan of an embryo along the cranio-caudal (head-tail) axis. Each gene has a specific region of the developing embryo that it regulates through its expression. These are conserved across the animal kingdom and represent a common ancestry.

It is interesting to note that if apoptosis is impaired this can lead to uncontrolled cell proliferation or death, and therefore, defects in development or even cancer could occur.

2.E.2: Timing and coordination of physiological events are regulated by multiple mechanisms.

The complex functions of organisms tend not to occur randomly. This would be an inefficient use of energy with consequential outcomes on survival. Instead well-timed physiology allows species to produce dependable and repeatable responses to changes in their development and their environmental cues. Specialized chemicals that have evolved to help signal physiological responses are called **pheromones**, which are secreted by a plant or an animal. In insects, for example, they influence the behavior or development of others of the same species, often functioning as an attractant of the opposite sex or driving adaptation to environmental changes. These environmental responses of pheromones can influence mating activity. In many moth species for example, female pheromone production and release and male pheromone responsiveness all show diurnal variations in many species, meaning that the moths activity at certain times at night can be related to the production or release of the pheromone. These temporal responses to day and night are chemically driven circadian rhythms, which act like clocks to synchronize physiology and optimize mating success. In primates, pheromones promote visual displays during the reproductive cycle in baboons. This ensures that males have a clear signal when the female is at peak fertility and increases the chance of finding a successful mate.

Induction of transcription factors during development results in sequential gene expression.

In plants, physiological events involve interactions between environmental stimuli and internal molecular signals.

In animals, fungi, protists and bacteria, internal and external signals regulate a variety of physiological responses that synchronize with environmental cycles and cues.

2.E.3: Timing and coordination of behavior are regulated by various mechanisms and are important in natural selection.

As we previously discussed, circadian rhythms are often directed by environmental cues that drive pheromone responses. This is beneficial not only in reproduction, but also survival. Changes in day length and average ambient temperature can signal to many animals the changes in the seasons. In mammals and insects that live in cold weather climates, hibernation or diapause (respectively) are important for the animal's survival through the limited resources of winter. Fat stores provide the animal with the metabolic energy it needs to survive through the cold, reducing its overall metabolic rate to ensure it has enough to make it through the winter. Likewise, when climates are warm and there are low resources available, some organisms will undergo aestivation, which conserve metabolic function and conserves water. Similar signals can even delay development to ensure that organisms (that would otherwise be competing for the same resources) reduce their time of competition. For example, some aphid species that feed on trees emerge from winter at different times of the year, reducing the likelihood of competing for the same tree at the same time and increasing their overall success.

Responses to information and communication of information are vital to natural selection.

Keywords

Cellular respiration
Oxidation-reduction reaction
Glycolysis
    Pyruvate
    Acetyl coA
    Link Reaction
Krebs cycle
Electron transport chain
Chemiosmosis
ATP Synthase
Anaerobic respiration – fermentation
Photosynthesis
    Autotroph
    Heterotroph
    Chloroplast
    Chlorophyll a
    Chlorophyll b
    Carotenoid
    Photosystem I
    Photosystem II
Light-dependent reactions
    Cyclic photophosporylation
    Non-cyclic photophosphorylation
Calvin cycle
    Carbon fixation
    Ribulose bisphosphate
      carboxylase
Photorespiration
CAM — crassulacean acid metabolism
Ectothermy
Endothermy
ATP
Properties of water
    Cohesion
    Adhesion
    Expansion
Phospholipid bilayer
Hydrophobic

Hydrophilic
Fluid mosaic model
Selective permeability
    Aquaporins
    Passive transport
    Simple diffusion
    Osmosis
    Active transport
    Channel protein
    Carrier protein
Operon
    Lac operon
    Trp operon
Positive /negative feedback
    mechanisms
    Phagocytosis
PEP carboxylase
Photoperiod
Phototropism
Taxis
Kinesis
Circadian Rhythm
    Diurnal
    Nocturnal
Biotic
Abiotic
Homeostasis
Non-specific immunity
Specific Immunity
    Antigen
    Antibody
    Major histocompatibility complex
      (MHC)
Morphogenesis
Apoptosis
Homeotic genes
Pheromones

Summary _____

Photosynthesis and chemosynthesis are used by autotrophs to capture free energy that is present in sunlight and inorganic chemicals, respectively. During cellular respiration and fermentation, energy is harvested from sugars to drive metabolic pathways.

Matter is exchanged between cells and organisms and the environment. In addition to water and nutrients, carbon and oxygen are required to synthesize new molecules that will be incorporated into carbohydrates, proteins, nucleic acids, and fats.

Membranes enable the internal environment to be separated from the external environment, and allows the movement of molecules in several manners: passive transport (osmosis and diffusion), and active transport.

Long-term survival of populations is achieved by the regulated timing and coordination of developmental, physiological, and behavioral events. These can include feedback mechanisms, changes in the environment of the biological system (the availability of resources), and homeostatic mechanisms.

Plants and animals also have defense mechanisms to combat disruptions to homeostasis.

Chapter 2 Quiz

Questions 1 and 2 relate to the data table below:

Oxygen Production	
Distance From Light (cm)	Bubbles Produced per Minute
10	39
20	22
30	8
40	5

1. **Based upon the findings from this study, what conclusion can be drawn about why primary producers have to live within the uppermost regions of the oceans, ponds, and lakes?**

   A. They have to live close to the surface in order to hunt for food.

   B. They have to live close to the surface in order to obtain the light they need for photosynthesis.

   C. They have to live close to the surface in order to produce enough energy to reproduce.

   D. They have to live close to the surface in order to excrete toxic levels of carbon dioxide.

2. **Which of the following processes is supported by the data?**

   A. $H2O + CO2 \xrightarrow{\text{light}} sugars + O2$

   B. $sugars + O2 \xrightarrow{\text{light}} H2O + CO2$

   C. $H2O + O2 \xrightarrow{\text{light}} sugars + CO2$

   D. $O2 + CO2 \xrightarrow{\text{light}} CO2 + sugars$

3. What can be concluded about the relationship between light intensity and the rate of photosynthesis, as seen in the graph?

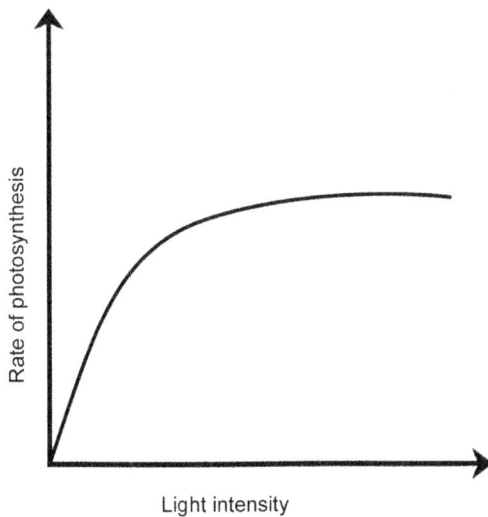

Light intensity

A. If this plant were given more light, the rate of photosynthesis will continue to increase.

B. Photosynthesis is not dependent upon light intensity, but rather the temperature of the air.

C. The rate of photosynthesis increases to a point of saturation regardless of how much light it is given.

D. If light intensity were decreased, the rate of photosynthesis would continue to rise due to other external factors.

4. What is the main difference between fermentation and glycolysis?

A. Fermentation breaks down sugars, while glycolysis breaks down proteins.

B. Fermentation requires an anaerobic environment, while glycolysis requires an aerobic environment.

C. Fermentation produces six molecules of ATP, while glycolysis produces only two.

D. Fermentation releases carbon dioxide into the atmosphere, while glycolysis releases oxygen.

5. **During the fall, many deciduous trees change color. For example, sugar maples of New England tend to go from green to bright oranges, yellows, and reds. What is the most likely explanation for this?**

   A. The absorption spectrum of chlorophyll changes during the fall to include green wavelengths.

   B. The trees increase their production of these pigments to adapt to the different amounts of light.

   C. There is a reduction in the production of green chlorophyll, so the masked pigments become visible.

   D. There are more orange, yellow, and red wavelengths found in the light during the fall than in the spring and summer.

6. **What would most likely happen to a plant cell that was placed into an isotonic solution?**

   A. It would become turgid.

   B. It would become flaccid.

   C. It would swell and lyse.

   D. It would undergo apoptosis.

7. **Water is an essential molecule needed by all life on Earth. Which property of water makes it so useful?**

   A. It is a polar ionic molecule.

   B. It is a polar covalent molecule.

   C. It is a nonpolar ionic molecule.

   D. It is a nonpolar covalent molecule.

8. **Which is of the following statements best describes why the Na^+K^+ sodium potassium pump is considered an active transporter?**

   A. Active transport is a form of facilitated diffusion.

   B. Ions are transported by the effects of gradients.

   C. Ions are transported by the effects of ATP.

   D. ATP is produced by the Na^+K^+ pump.

9. Scientists think they have discovered a new bacterium that has an operon to control the digestion of a novel sugar. This operon acts similarly to the *trp* operon. What would be the expected response in the presence of a large amount of the novel sugar?

   A. The repressor will be activated, and gene expression will be turned on.

   B. The repressor will be inactivated, and gene expression will be turned on.

   C. A positive feedback mechanism will occur.

   D. A negative feedback mechanism will occur.

10. In a lab experiment, researchers provide several plants a greenhouse and supplied them with radioactively labeled H_2O. What would the researchers observe when the plants perform photosynthesis?

   A. The plants will release radioactive CO_2.

   B. The plants will release radioactive O_2.

   C. The plants will produce glucose.

   D. The plants will utilize CO_2.

Chapter 2 Quiz Answer Key

1. **Answer: B.** Primary producers are highly dependent upon light to make food for themselves. Through photosynthesis, they take in carbon dioxide and water and produce sugars, which are used for energy. A byproduct of this process is oxygen, as represented by the bubbles in the data. The closer the producer is to the light source, the more photosynthesis takes place, and the more bubbles will be produced.

2. **Answer: A.** The process supported by the data is photosynthesis. In this biochemical reaction, producers take in water and carbon dioxide and convert them into sugars. Oxygen is a byproduct.

3. **Answer: C.** The graph shows that the rate of photosynthesis and light intensity are related, but only to a point. As more light is provided, the rate increases. However, once it reaches a saturation point, it does not matter how much light is given. The chloroplasts cannot work any harder.

4. **Answer: B.** Fermentation takes place in the absence of oxygen. For example, if you were making wine, you would place the grapes and yeast in an airtight barrel for the process to occur. Glycolysis needs oxygen in order to progress.

5. **Answer: C.** Chlorophyll is the pigment that makes leaves green. During the fall there is less light available, so chlorophyll production decreases. As a result, the green color of leaves disappears, and the masked pigments become visible.

6. **Answer: B.** Plant cells have a cell wall, which prevents lysing (rupture of the cell membrane). Isotonic environments do not allow for swelling or turgidity, so plant cells become flaccid.

7. **Answer: B.** Water is a polar covalent molecule. This means that the atoms that bond together to form it are sharing electrons in their outermost shells. It also means that after the atoms have bonded, there is a positive side and a negative side to the molecule.

8. **Answer: C.** ATP drives transport of ions against their concentration gradient in the Na^+K^+ pump in a process that consumes ATP. Facilitated diffusion is a type of passive transport, not active transport.

9. **Answer: D.** The *trp* operon is always activated. Therefore, in this case, high levels of the sugar will bind to the repressor, preventing the synthesis of more sugar – a prime example of negative feedback.

10. **Answer: B.** During photosynthesis, oxygen is released after water is split. Since carbon dioxide is utilized in the dark and produces glucose, the other answers are wrong in this case because the reaction is occurring in the presence of light.

Chapter 3.
Big Idea 3: Information

Big Idea 3: Information: Living systems store, retrieve, transmit and respond to information essential to life processes.

What you will learn from this chapter:

- Production and function of nucleic acids: DNA and RNA
- Differences between mitosis and meiosis
- How genetic variation occurs
- Signal transduction and gene expression
- Endocrine system and hormones
- Nervous system responses

Introduction

What is the basis of life? You always hear that cells are the building blocks of life, but cells are made up of billions of smaller molecules including DNA and RNA, just to name a few. Each of these molecules has specific characteristics allowing them to contribute to the function and activity of the cell, and the organism as a whole. How do they do it? Does dysfunction in DNA, RNA, or protein construction lead to diseases? Does the disease lead to dysfunction? Let's find out!

Enduring Understanding 3.A: Heritable information provides for continuity of life

3.A.1: DNA, and in some cases RNA, is the primary source of heritable information

Gregor Mendel and the dawn of genetics

Long before the discovery of DNA, and just as Darwin's ideas about natural selection were hitting the world with a bang, a parallel, quiet revolution was taking place in the garden plots of a Moravian monastery. Gregor Johann Mendel, a monk and physics teacher, was more interested in meteorology than in biology, but he was the son of farmers, a friend of agricultural biologists, and suspicious of the prevalent view of inheritance of biological traits. It was already known that heritable variation was responsible for the individuality of organisms; from wheat to dogs, breeders had mixed genetic material in order to develop organisms with desirable traits. However, the prevailing idea was that inheritance happened through a smooth mixing of parental genetic material; that genetic material was like paint of different colors that blended in the next generation. Mendel hypothesized that inheritance happened via discrete, individual traits rather than continuous qualities that mixed together. To test this, he conducted a systematic set of experiments, using plants that were closely related enough to allow the study of individual traits. From this work, he developed the two laws, now known as Mendel's Laws of Inheritance, that earned him, posthumously, the respect of the biological community as the father of genetics.

Since genes and DNA had not been discovered at the time, Mendel drew his conclusions purely on the basis of **phenotype**; that is, the outwardly expressed, physical traits that result from the underlying **genotype**, which we now know is encoded in DNA. On a biochemical level, phenotype is the result of proteins (or functional RNA)—in other words, **expressed genes**. Genotype is on the level of DNA, or **encoded genes**. Genes may be expressed—leading to a phenotype—or not expressed ("carried") and passed to a later generation that expresses them. If this is the case, the observed phenotype is quite different from the parents. (Phenotype can also be affected by the environment; for example, an organism may have a gene for tall stature, but be very short because of poor nutrition during development.)

DNA and RNA

A **nucleic acid** functions to encode, transmit, and express genetic information. A monomer, or a building block of nucleic acid is called a nucleotide. A nucleic acid is composed of a 5-carbon sugar, a phosphate group, and a nitrogen base. The two types of nucleic acids are deoxyribonucleic acid (DNA) and ribonucleic acid (RNA). DNA contains information for genes, and RNA contains information to generate proteins.

There are five bases that generate the code contained in DNA and RNA: adenine, thymine, cytosine, guanine, and uracil. Evolution has conserved base pairing so that cytosine pairs with guanine and adenine pairs with thymine. Uracil is found only in RNA, and replaces thymine to pair with adenine. Adenine and thymine (or uracil in RNA) are linked by two hydrogen bonds and cytosine and guanine are linked by three hydrogen bonds, making the bond between cytosine and guanine harder to break apart because of the greater number of

Phenotypes are determined through protein activities.

DNA and RNA molecules have structural similarities and differences that define function.

bonds between the bases. Adenine and guanine are purines that have a double ring structure; while cytosine, thymine, and uracil are pyrimidines that have a single ring structure. The structure and pairing of bases contributes to the symmetry of the double-stranded DNA molecule in a double helix, or twisted ladder, shape as seen below.

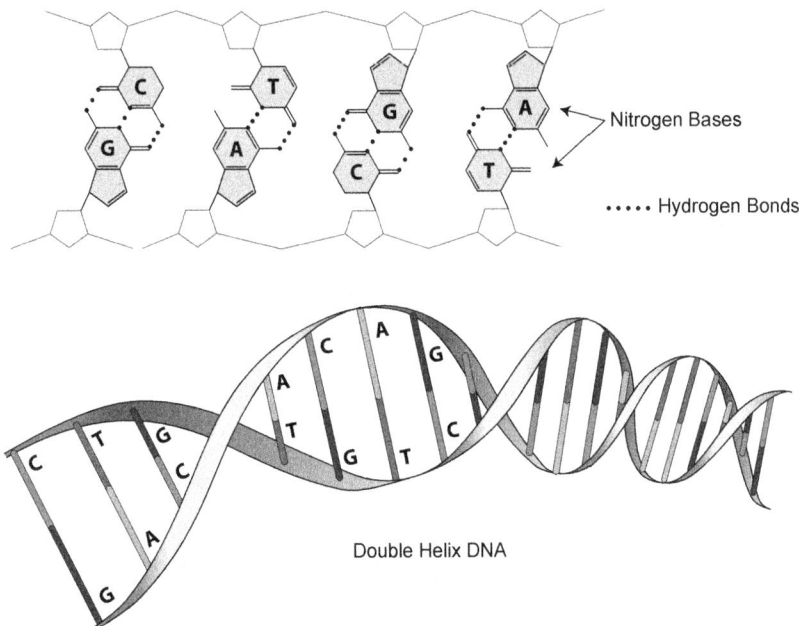

Double Helix DNA

An Illustration of DNA structure and pairing between bases in the DNA molecule.

The following provides a summary of nucleic acid composition:

	SUGAR	PHOSPHATE	BASES
DNA	Deoxyribose	Present	adenine, **thymine**, cytosine, guanine
RNA	Ribose	Present	adenine, **uracil**, cytosine, guanine

DNA replicates as a **semiconservative** process meaning the only one strand of the original DNA molecule is conserved and serves as a template for the new strand.

DNA replication occurs in a series of steps with the help of enzymes to complete the reactions:

1. Separation of the two strands of DNA, is facilitated by an enzyme called a **helicase**.
2. During the strand separation, it is necessary to reduce tension at the area of the separated strands, called the **replication fork**. **Topoisomerases** are enzymes that relieve the tension by nicking one strand and relaxing the supercoil.
3. To initiate DNA synthesis, **RNA polymerase** adds ribonucleotides to the DNA template. This short RNA-DNA hybrid is called a **primer**.
4. The single-stranded DNA can then have nucleotides added using DNA polymerases, which normally occurs in the 5' → 3' direction.
5. DNA replication is bi-directional: the **leading strand** is continuously synthesized in the direction that the replication fork is moving, and the lagging strand is discontinuously synthesized in the opposite direction. As the replication fork

proceeds, new primer is added to the **lagging strand** and it is synthesized discontinuously in small fragments called **Okazaki fragments**.

6. The two strands are bonded together by reactions catalyzed by an enzyme called **DNA ligase** which seals the nicks in the strands, resulting in the final product, a double-stranded segment of DNA.

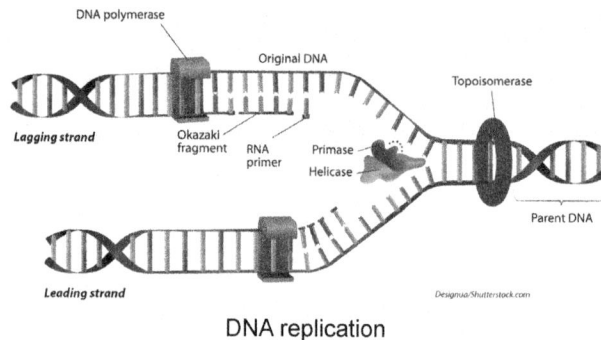

DNA replication

Chromosomal replication in bacteria is similar to eukaryotic DNA replication. A **plasmid** is a small ring of DNA that carries accessory genes separate from those of the bacterial chromosome. Most plasmids in Gram-negative bacteria undergo bi-directional replication, although some undergo unidirectional replication because of their small size. Plasmids in Gram-positive bacteria replicate by a process called rolling circle mechanism.

Some plasmids can transfer themselves (and therefore their genetic information) to another cell through a process called **conjugation**. Conjugation requires cell-to-cell contact. The sex pilus of the donor cell attaches to the recipient cell. Once contact has been established, the transfer of DNA occurs by a rolling circle mechanism.

Protein synthesis

Proteins are synthesized through the process of translation. Three major classes of RNA are required:

- **Messenger RNA** contains information for translation.
- **Ribosomal RNA** is a structural component of the ribosome.
- **Transfer RNA** carries amino acids to the ribosome for protein synthesis.

The first step of protein synthesis is **transcription**, when the DNA molecule is copied into an RNA molecule (mRNA). This process is similar in prokaryotes and eukaryotes and has three steps:

1. **Initiation** begins at the **promoter** – which is a specific sequence of the double-stranded DNA molecule. The double-stranded DNA opens up and RNA polymerase begins transcription in the $5' \rightarrow 3'$ direction, on the **sense strand**. The strand that is not used, is called **antisense**. The ribonucleotides uracil and cytosine are paired to the deoxyribonucleotides adenine and guanine respectively, to get a complementary mRNA segment.

2. **Elongation** is the synthesis of the mRNA strand in the $5' \rightarrow 3'$ direction. The new mRNA rapidly separates from the DNA template and the complementary DNA strands pair together.

Genetic information flows from a sequence of nucleotides in a gene to a sequence of amino acids in a protein.

3. **Termination** of transcription occurs at the end of a gene. The mRNA is separated from the DNA template at a specific sequence to direct termination.

In eukaryotes, mRNA goes through **post-transcriptional processing** before continuing to translation. There are three steps of processing:

1. **5' capping** – The addition of a GTP cap to protect the 5' end from degradation, serving as the site where ribosomes bind to the mRNA for translation.
2. **3' polyadenylation** – The addition of 100–300 adenines to the free 3' end of mRNA resulting in a poly-A-tail that is important for the stability of the mRNA as well as for nuclear export..
3. **Intron removal** – The removal of non-coding introns and the splicing together of coding exons to form the mature mRNA.

Translation is the process in which the mRNA sequence is used to synthesize a polypeptide. The mRNA sequence determines the amino acid sequence of a protein by following a pattern called the **genetic code**. The genetic code consists of triplet nucleotide combinations called codons, which encode 20 different amino acids—the building blocks of protein. They are joined together by peptide bonds to form a polypeptide chain.

Ribosomes are parts of the cell that are the site of translation. The process of translation also occurs in three steps which is initiated by binding of methylated tRNA to the ribosome at the start codon, AUG. This complex then binds to the 5' cap of the mRNA. Next, tRNAs begin elongation by carrying the amino acid to the ribosome according to the mRNA sequence. tRNA is very specific – it only accepts one of the 20 amino acids that correspond to the anticodon. The anticodon is complementary to the codon. For example, using the codon sequence below:

the mRNA reads   A U G / G A G / C A U / G C U
the anticodons are   U A C / C U C / G U A / C G A

The final step is termination, which occurs when the ribosome reaches any one of the three stop codons: UAA, UAG, or UGA. The newly formed polypeptide then undergoes posttranslational modification to alter or remove portions of the polypeptide.

Some viruses have RNA genomes, and therefore their genotype is encoded in RNA. They use an enzyme called reverse transcriptase to convert their RNA genome into a DNA copy that is readable by the infected cell's machinery. Biologists think that the early life forms originally contained only RNA; this would be possible because RNA can both encode genes and act as a functional molecule, since it can fold into a unique 3D tertiary structure as proteins do. The most well-known examples of functional RNA molecules are transfer RNAs and ribosomal RNAs.

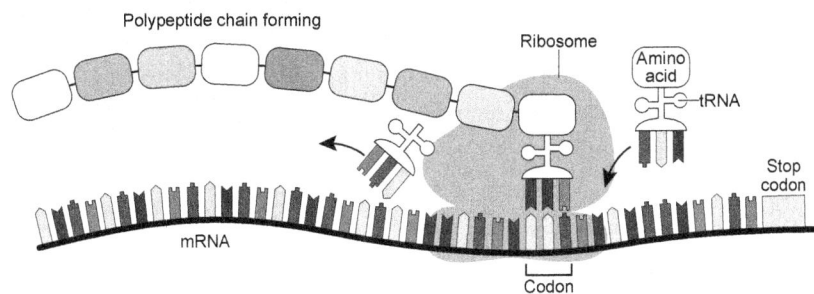

Genetic engineering

Modifying the genetic composition of an organism by artificial means is called genetic engineering, which has been helpful for the treatment of some diseases. In its simplest form, genetic engineering requires enzymes to cut DNA, a vector, and a host organism in which to place the recombinant DNA. A **restriction enzyme** is a bacterial enzyme that cuts foreign DNA in specific locations. The restriction fragment that results can be inserted into a bacterial plasmid (vector). Other vectors that may be used include viruses and bacteriophages. The splicing of restriction fragments into a plasmid results in a recombinant plasmid. This recombinant plasmid can then be placed in a host cell, usually a bacterial cell, for replication. For example, insulin has been produced in bacteria by gene-splicing techniques. The insulin produced in genetically engineered bacteria is chemically identical to that made in the pancreas. Insulin treatment helps control diabetes for millions of people who suffer from the disease, providing evidence that genetic engineering is beneficial for ameliorating disease.

The use of recombinant DNA also provides a means to transfer genes among species. This opens the door for cloning specific genes of interest. Hybridization can be used to find a gene of interest. A probe is a molecule complementary to the sequence of a gene of interest. The probe, once annealed to the gene, can be detected by labeling with a radioactive isotope or a fluorescent tag. This allows one to track the gene or determine the efficiency of the treatment.

Gene therapy attempts to treat disorders by employing techniques of genetic engineering, most commonly by inserting a functional or therapeutic gene into a cell in order to replace a dysfunctional protein. Gene therapy has allowed doctors and scientists to introduce a normal allele of an enzyme that is missing in patients afflicted with certain diseases.

The use of DNA probes and the **polymerase chain reaction** (PCR) has enabled scientists to identify and detect elusive pathogens, in addition to performing multiple procedures on small amounts of DNA. PCR is a technique in which a piece of DNA can be amplified into billions of copies within a few hours. This is now a basic, standard laboratory procedure that makes it possible to diagnosis genetic diseases before the onset of symptoms. This process requires a primer to specify the segment to be copied, and an enzyme (usually taq polymerase) to amplify the DNA.

A method used to visualize DNA fragments produced in PCR is **gel electrophoresis**. Electrophoresis separates DNA or protein by size or electrical charge. The DNA runs towards the positive charge and the DNA fragments separate by size. The gel is treated with a DNA-binding dye that fluoresces under ultraviolet light and the sizes of the DNA fragments, or PCR products, can be measured against a DNA ladder containing fragments of known sizes. An image of PCR products visualized by gel electrophoresis is illustrated here:

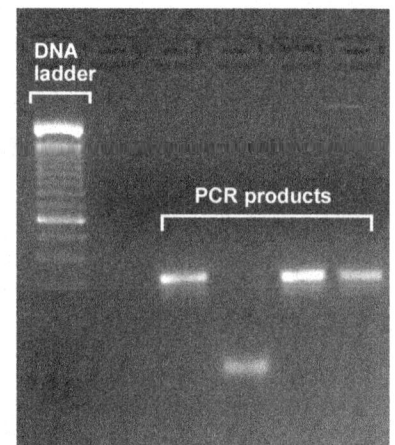

Image of DNA bands on agarose gel

Mapping Genomes

The human genome project was a large undertaking to map and sequence the three billion nucleotides in the human genome, and to identify all of the genes encoded within. The project was launched in 1986 and an outline of the genome was finished in 2000 through international collaboration. In May 2006, the sequence of the last chromosome was published! While the map and sequencing are complete, scientists are still studying the functions and regulation of all the genes. The genome of other mammals has also been successfully decoded, and scientists believe that this information will allow us to better understand evolution and develop treatments for diseases.

3.A.2: In eukaryotes, heritable information is passed to the next generation via processes that include the cell cycle and mitosis or meiosis plus fertilization.

The purpose of cell division is to provide growth and repair in body (somatic) cells and to replenish or create sex cells for reproduction. There are two forms of cell division: mitosis and meiosis. **Mitosis** is the division of somatic cells and meiosis is the division of sex cells (eggs and sperm). It is important to note that the number of sets of chromosomes contained in a cell depends on the type of cell reproduction. A **diploid** cell has two sets of chromosomes, and is a result of mitosis. A **haploid** cell has only one set of chromosomes, and is a result of meiosis.

Cell cycle

The life cycle of each cell is the time between one cell division to the next cell division. When a cell is not dividing, it needs time to produce the necessary materials for division. This period is called interphase, and it occurs in three stages:

- G_1 (growth) period, when the cell grows and produces enzymes necessary for DNA replication
- S (synthesis) period, when the cell copies the chromosomes—creating new DNA— in preparation for the mitotic phase.
- G_2 (growth) period, when the cell makes new proteins and organelles in preparation for cell division.

Mitosis and cytokinesis divide the nucleus and cytoplasm, respectively. The mitotic phase is the shortest phase of the cell cycle.

The cell cycle is a complex set of stages that is highly regulated with checkpoints, which determine the ultimate fate of the cell.

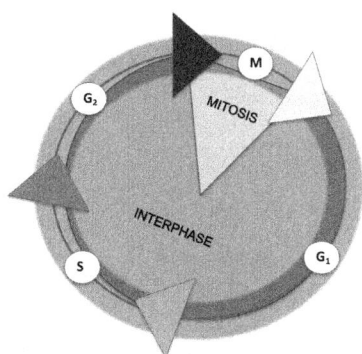

Mitosis

The mitotic phase is a continuum of change, although we divide it into distinct stages: prophase, metaphase, anaphase, and telophase.

1. During **prophase**, the chromatin condenses to become visible chromosomes, the nucleolus disappears, and the nuclear membrane breaks apart. Next, centrioles will align themselves at either pole of the cell and send out an array of microtubules to form the spindle fibers that will pull the chromosomes apart. Then, the nuclear membrane fragments and allows the spindle microtubules to interact with the chromosomes.

2. In **metaphase**, the centromeres (parts of the chromosomes that link the chromosome pairs – sister chromatids) allow the chromosomes to align along the central plane of the cell, attached to the spindle fibers by kinetochores.

3. During **anaphase**, the centromeres split in half and homologous chromosomes separate. The chromosomes are pulled to the poles of the cell, with identical sets at either end.

4. In **telophase**, two nuclei form with a full set of DNA that is identical to the parent cell. The nucleoli become visible and the nuclear membrane reassembles.

5. Finally, **cytokinesis**, or division of the cytoplasm and organelles, occurs. A cell plate forms in plant cells and a cleavage furrow forms in animal cells. The cell pinches into two cells.

Below is a diagram of the actions and structural changes during mitosis.

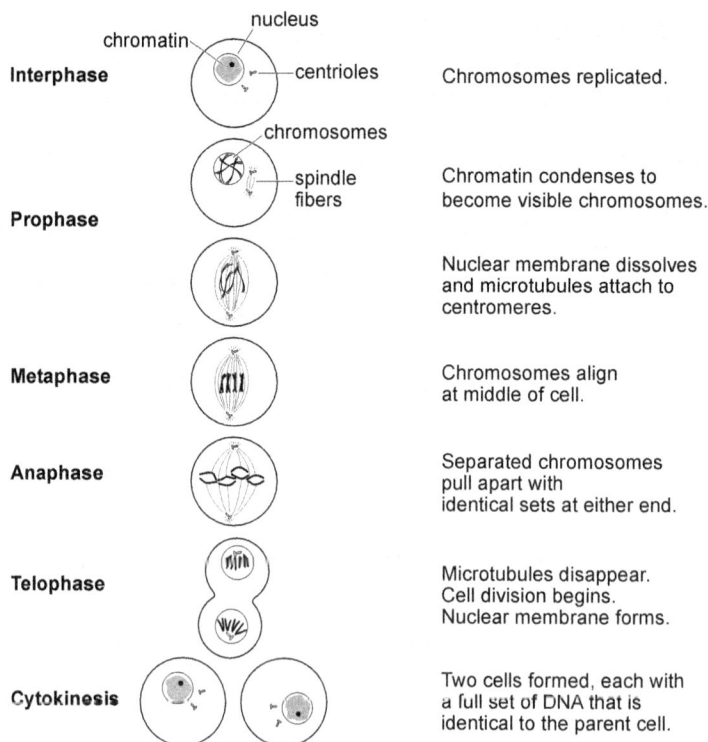

Damage to the DNA of a cell can result in the inability of a cell to divide. If the damage cannot be repaired, the life of the cell can end in a programmed cell death called **apoptosis**, as we briefly discussed earlier. Apoptosis is still not fully understood, though some believe it evolved as a mechanism to prevent the uncontrolled cell division that can lead to cancer. The more times a cell divides, the more chances it has to accumulate errors in its DNA. Therefore, it seems logical that a mechanism would have evolved to prevent damaged cells from continuing to divide. Scientists currently believe that telomeres (the highly repetitive DNA on the ends of chromosomes), which appear to shorten during each cell division, may play a regulatory role in limiting the number of cell divisions and triggering senescence. This has implications on understanding the aging process, and research is underway to determine if aging can be slowed.

Meiosis

Meiosis is similar to mitosis, but there are two consecutive cell divisions, meiosis I and meiosis II, in order to reduce the chromosome number by half. This way, when the sperm and egg join during fertilization, the result is an organism with two sets of chromosomes.

Similar to mitosis, meiosis is preceded by an interphase during which the chromosomes are replicated. The steps of meiosis are as follows:

1. **Prophase I:** The replicated chromosomes pair in a process called **synapsis:** Two sets of chromosomes form a **tetrad** and **cross over,** exchanging segments of genetic material between the pair of chromosomes, also called homologues, to further increase diversity.
2. **Metaphase I:** The homologous pairs attach to spindle fibers after lining up in the middle of the cell.
3. **Anaphase I:** The tetrad separates and one of each pair of chromosomes moves to opposite poles of the cell.
4. **Telophase I:** The homologous chromosome pairs continue to separate. Each pole now has a haploid chromosome set. Telophase I occurs simultaneously with cytokinesis.
5. **Prophase II:** A spindle apparatus forms and the chromosomes condense.
6. **Metaphase II:** Chromosomes line up in the center of cell individually, not as a pair.
7. **Anaphase II:** The separated chromosomes move to opposite ends of the cell.
8. **Telophase II:** The nuclear membrane forms around the set of chromosomes and cytokinesis occurs, resulting in four haploid daughter cells.

Meiosis, a reduction division, followed by fertilization ensures genetic diversity in sexually reproducing organisms.

Interphase		Chromosomes replicated.
Prophase I		Pairs of homologous chromosomes come together in tetrads and crossover.
Metaphase I		Homologous pairs attach to spindle fibers after lining up in the middle of the cell.
Anaphase I		Tetrads separate and one of each pair of chromosomes moves to opposite poles of the cell.
Telophase I		Haploid sets of chromosomes arrive at the poles. Cytokinesis creates two daughter cells.
Prophase II		Spindle apparatus forms and the chromosomes condense. Nuclear membrane dissolves.
Metaphase II		Chromosomes line up in center of cell individually, not as a pair
Anaphase II		Separated chromosomes move to opposite ends of the cell.
Telophase II		Microtubules disappear. Cell division begins. Nuclear membrane forms.
Cytokinesis		Four haploid cells are produced.

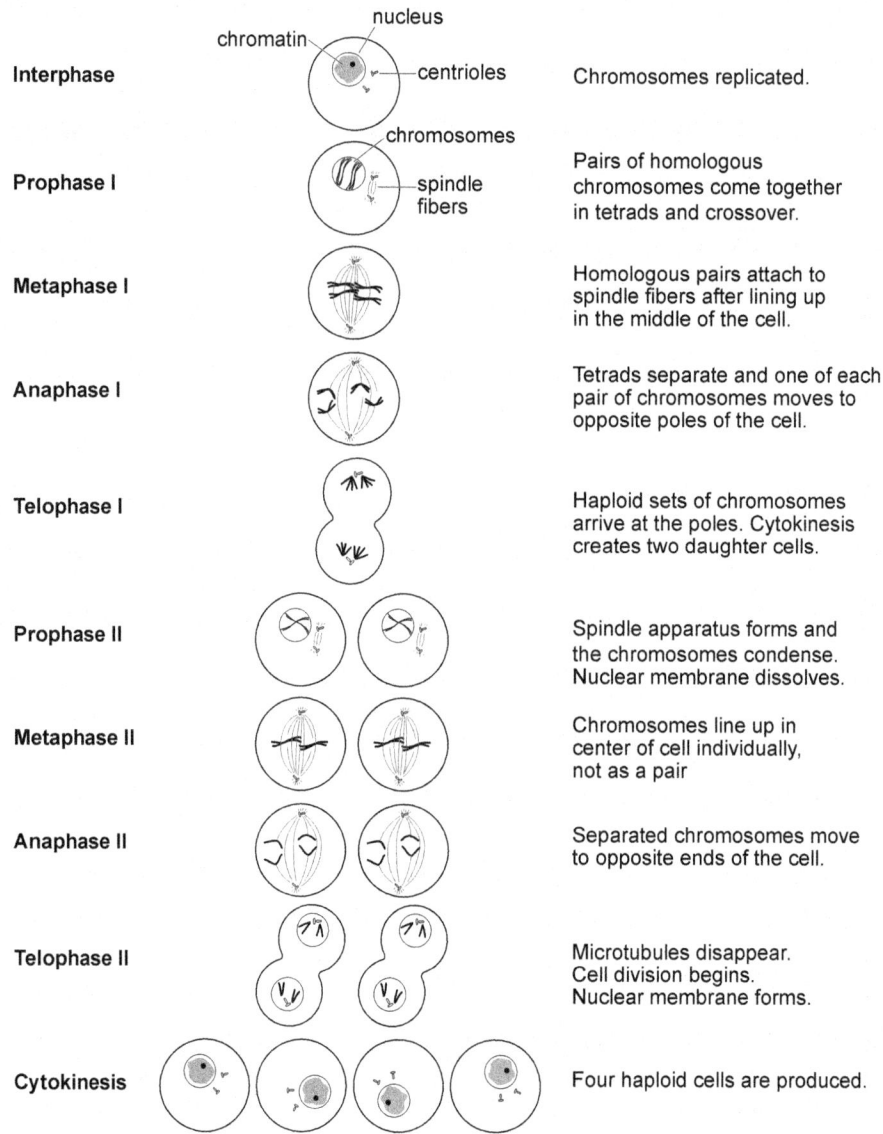

The stages of meiosis

To summarize the differences and similarities between mitosis and meiosis, we have provided the Venn diagram below:

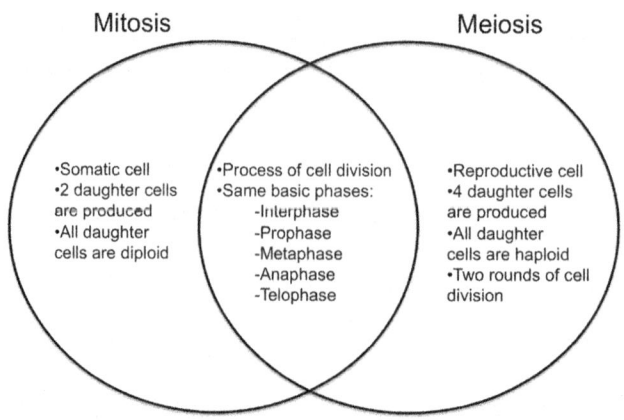

Mitosis — **Meiosis**

Mitosis:
•Somatic cell
•2 daughter cells are produced
•All daughter cells are diploid

Intersection:
•Process of cell division
•Same basic phases:
-Interphase
-Prophase
-Metaphase
-Anaphase
-Telophase

Meiosis:
•Reproductive cell
•4 daughter cells are produced
•All daughter cells are haploid
•Two rounds of cell division

3.A.3: The chromosomal basis of inheritance provides an understanding of the pattern of passage (transmission) of genes from parent to offspring

Mendel's Law of Segregation

Mendel worked with pure-bred pea plants that exhibited seven easily-observed binary phenotypic traits, known as **alleles.** Each allele took the form of one of two clearly distinguishable phenotypes, such as round vs. wrinkled seeds. He bred true-breeding (purebred) plants with plants of different phenotypes, and looked at the resulting offspring in the first and second generations. He found that the traits from the two parents did not blend, but instead seemed to be inherited independently, with one trait "dominating" the other. His results are summarized in the following table:

Trait	F1 phenotype results	F2 phenotype results	F2 phenotype ratio	Conclusion
Seeds: round vs. wrinkled	All round	5474 round, 1850 wrinkled	2.96 to 1	round seeds = dominant **traits do not blend**
Seeds: yellow vs. green	All yellow	6022 yellow, 2001 green	3.01 to 1	yellow seeds = dominant **traits do not blend**
Pods: inflated vs. constricted (smooth vs. bumpy)	All inflated	882 inflated, 299 constricted	2.95 to 1	inflated pods = dominant **traits do not blend**
Pods: green vs. yellow	All green	428 green, 152 yellow	2.82 to 1	green pods = dominant **traits do not blend**
Flowers: purple vs. white	All purple	705 purple, 224 white	3.15 to 1	purple flowers = dominant **traits do not blend**
Flower position: axial vs. terminal	All axial	651 axial, 207 terminal	3.14 to 1	axial flowers = dominant **traits do not blend**
Stem length: tall vs. dwarf	All tall	787 tall, 277 dwarf	2.84 to 1	tall plants = dominant **traits do not blend**

Mendel's data showed that the traits did not blend (tall plants breeding with dwarf plants did not produce medium-sized plants, for example) and one trait did not eliminate or change the other one (recessive factors were still there, since the traits re-emerged in the F2 generation.) Instead, each of the two "factors" underlying each allele retained its nature, more like different-colored marbles being put in a bag than different-colored paints being mixed together. Each parent plant gave one or the other of these two "factors" to the offspring, but not a blend of the two. This is known as the **Law of Segregation**. When Mendel bred one true-breeding plant with another, the next (F1) generation of plants carried both of the "factors" (now known as **genes**)—one from each parent. If we designate them as DR (for dominant and recessive), the F1 generation was 100% DR in genotype, but since the dominant gene gives rise to the phenotype, they all looked like the

Segregation and independent assortment of chromosomes result in genetic variation.

DD parent; none of them looked like the RR parent. In the F2 generation, these factors were again redistributed, so that ¼ of the offspring inherited dominant genes from both parents (DD), ¼ inherited a dominant gene from parent 1 and a recessive gene from parent 2 (DR), ¼ inherited a recessive gene from parent 1 and a dominant gene from parent 2 (RD– the same as DR), and ¼ inherited recessive genes from both parents (RR). The resulting ratio of 25% DD, 50% DR, and 25% RR was expressed phenotypically as 75% dominant and 25% recessive; 75% looked like the DD grandparent and 25% looked like the RR grandparent. In other words, plants with DD or DR genotypes had the dominant phenotype, while plants with the RR genotype had the recessive phenotype. Organisms with identical alleles—two copies of the same trait (DD or RR)—are known as **homozygous**, while those that carry different alleles (DR) are known as **heterozygous**.

The Law of Segregation states that only one of two possible alleles from each parent will be passed to the offspring. If the two alleles differ, then one is fully expressed in the organism's appearance (the dominant allele) and the other has no noticeable effect on appearance (the recessive allele). The two alleles for each trait segregate into different gametes. A **Punnett square** can be used to illustrate the law of segregation. In a Punnett square, one parent's genes are put at the top of the box and the other parent's on the side. Genes combine in the squares just like numbers are added in addition tables. Traditionally, a trait is given a letter designation; the dominant allele is designated by this letter in uppercase and the recessive allele is designated by the same letter in lowercase. This Punnett square shows the result of the cross of two F$_1$ hybrids, each of which have genotypes Pp:

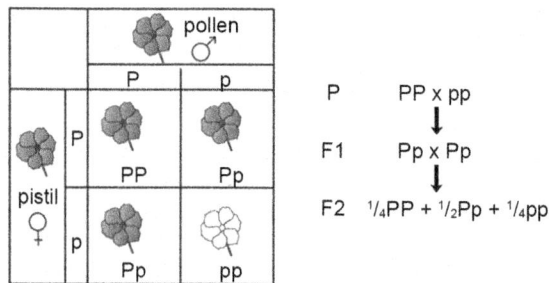

This cross results in a 1:2:1 ratio of F2 offspring. Here, the P is the dominant allele and the *p* is the recessive allele. The F1 cross produces three offspring expressing the dominant allele (one *PP* and two *Pp*) and one offspring expressing the recessive allele (*pp*). PP plants and pp plants are both homozygous, and Pp plants are heterozygous.

*A word about probability: The results of a breeding experiment like this are subject to probability. If you rolled a 4-sided die, you would expect to roll a "4" ¼ of the time, but in reality, you may end up rolling a "4" half the time. The more times you roll the dice, the closer the total number of "4" rolls will be to ¼ of the total number of rolls, but even with a large number of rolls, it will probably not be exactly 25%. The same goes for experimental results. It is therefore important to use (1) as large a number of subjects/trials as possible; (2) a careful experimental design to eliminate random factors that could skew results, called **confounding factors** (Mendel used true-breeding, simple plants for this reason); and (3) calculate indices of variation, such as standard deviation, to estimate the likelihood that a result is due to chance alone vs. a true effect.*

Mendel chose his experimental plants because their simplicity allowed him to look at inheritance mathematically and thereby deduce the discrete nature of genetic phenomena. The "real world" of most genomes, however, is much more complex. In Mendel's experiments, the F1 generation had either purple or white flowers. This is an example of **complete dominance**. However, **incomplete dominance** is when the F1 generation results in an appearance somewhere between the two parents. For example, red flowers crossed with white flowers resulting in an F1 generation with pink flowers. The F2 generation will show three phenotypes, red, pink, and white, in a ratio of 1:2:1, respectively. In **codominance**, both genes are expressed fully. This is in part why, currently, the principles of segregation and independent assortment are considered Mendelian laws, but the principle of dominance, formerly considered a law by many, is considered a principle that only applies to some traits.

ABO blood grouping provides an example of both complete dominance and codominance. A and B are genes that express different types of glycoprotein antigen on the surfaces of blood cells. These antigens are like little "flags" that tell the immune system that a blood cell comes from a person's own body and therefore should not be attacked. If a person with type-A blood is injected with an infusion of type-B blood, his or her immune system will attack the new blood, with its "enemy flags", as if it were a pathogenic organism invading the body. Whereas types A and B can be seen as having different "flags", type O blood has no "flag", and type AB blood has both "flags".

If someone receives a type A gene from one parent and a type O gene from the other, his blood will have type A antigens and he will therefore have type A blood; type A is dominant and type O is recessive. The same will occur with type B blood; if a person has the genotype BO, she will have type B blood, the same phenotype as a person with a BB genotype. Children with two O genes, however, will have type O blood; people with type O blood are all homozygous recessive. A and B are each dominant over O, but what if a child receives an A gene from one parent and a B gene from the other? Such a child will have AB blood—both antigens will be expressed. AB blood type is an example of codominance. In summary, type A blood may have the genotypes of AA or AO, type B blood may have the genotypes of BB or BO, type AB blood always has the genotype AB, and type O blood always has the genotype OO. Blood type shows that complete dominance and codominance are not different types of inheritance; the degree of dominance is simply a function of the nature of the protein that each gene encodes.

Mendel's Law of Independent Assortment

Alongside his observation that traits did not blend, Mendel observed that traits assorted independently of other traits. In other words, an F1 plant could be tall and have purple flowers, but the F2 generation could be tall with purple flowers, tall with white flowers, dwarf with purple flowers, or dwarf with white flowers. Plant height did not affect flower color. The **Law of Independent Assortment** states that alleles assort independently of each other. This is illustrated in a Punnett square of a dihybrid cross, in which two characters are explored. Two of the seven characteristics that Mendel studied were seed shape and color. Yellow is the dominant seed color (Y) and green is the recessive color (y). The dominant seed shape is round (R) and the recessive shape is wrinkled (r). A cross between a plant with yellow round seeds ($YYRR$) and a plant with green wrinkled seeds ($yyrr$) produces an F1 generation with the genotype $YyRr$. Independent assortment of the $YyRr$ genotype yields four possible genotypes: YR, Yr, yR, and yr. Crossing the F1 generation would result in the production of F2 offspring with a 9:3:3:1 phenotypic ratio.

pistil ♀ \ pollen ♂	YR	Yr	yR	yr
YR	YYRR	YYRr	YyRR	YyRr
Yr	YYRr	YYrr	YyRr	Yyrr
yR	YyRR	YyRr	yyRR	yyRr
yr	YyRr	Yyrr	yyRr	yyrr

P   YYRR × yyrr
          ↓
F1   YyRr
          ↓
F2   1 - YYRR ⎫
      2 - YYRr ⎬ 9 yellow round
      2 - YyRR ⎪
      4 - YyRr ⎭

      1 - yyRR ⎫ 3 green round
      2 - yyRr ⎭

      1 - YYrr ⎫ 3 yellow wrinkled
      2 - Yyrr ⎭

      1 - yyrr ⎭ 1 green wrinkled

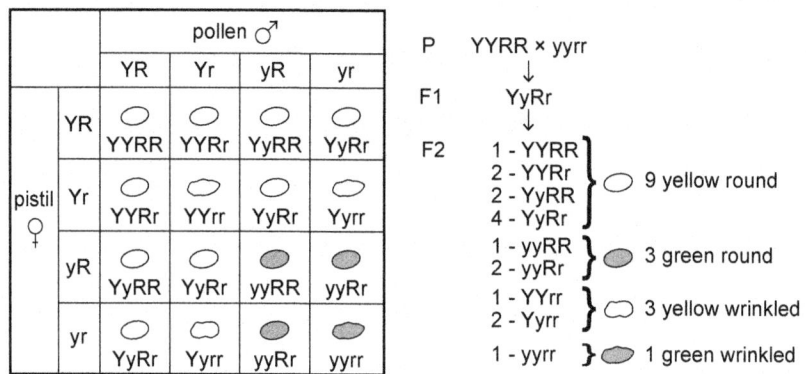

Sex-linked traits

On the surface, sex-linked traits may seem not to obey Mendel's laws, since they are not inherited in the expected 75%–25% ratio for dominant and recessive genes. However, this is not due to non-Mendelian inheritance but due to gene expression; the genes themselves obey the laws of segregation and independent assortment. Let's take a closer look at why the ratios of sex-linked genes are different from those of genes on non-sex chromosomes (**autosomal** genes).

The Y chromosome is very short, containing very few genes. Therefore, genes on the X chromosome that are recessive for females are dominant for males. This is another illustration of the fact that dominance is not a function of inheritance; it is a function of the gene expression products. For example, the *OPN1LW* gene encodes retinal proteins optimized to the red-orange part of the spectrum. It is located on the X chromosome. If a woman has a mutated copy of the *OPN1LW* gene, she will still be able to see red-orange colors properly as long as she has a normal (functional) copy of the *OPN1LW* gene on her other X chromosome. However, if a man has a mutated *OPN1LW* gene, he will have problems seeing the color red, because he does not have a second X chromosome to provide a functional copy of *OPN1LW*.

Consider a situation in which a father has a normal *OPN1LW* gene (let's call it R) and the mother is heterozygous, with one functional and one nonfunctional copy of the gene (Rr). The father, with his R gene, can see red just fine. The mother, who also has an R gene, can see red as well. However, let's consider their children. The girls will receive one X chromosome from their father and one X chromosome from their mother. Therefore, half of the girls will be RR and half will be Rr. All of the girls will be phenotypically normal, since they all have an R gene; R is dominant, so only girls with rr genomes will have trouble seeing red. The boys, however, will receive a Y chromosome from their father and an X chromosome from their mother. Since the gene is only found on the X chromosome, the boys will only have one copy. Half of them will receive R from their Rr mother, and half of them will receive r. The half with r will have trouble seeing the color red. In the next generation, if they mate with a woman who is RR, all of the color-blind boys will have sons with full color vision, since they will give their sons their Y chromosome. Their daughters however, will all inherit the r gene; they will see normally, but will have the genotype Rr, resulting in half of the following generation of sons being color-blind.

Sex-limited traits are the not the same as sex-linked traits. Sex-limited traits are only expressed in one sex, such a milk production (only in females) and male-pattern baldness (only in males).

The pedigree

The pattern of inheritance (monohybrid, dihybrid, sex-linked, and genes linked on the same chromosome) can often be predicted from data on the parent genotype/phenotype and/or the offspring phenotypes/genotypes. One common way to visualize transmission is a pedigree. A family pedigree is a collection of a family's history for a particular trait. Focusing on a trait of interest, the generations are mapped in a pedigree chart, similar to a family tree but with the alleles present. In a case where both parents have a particular trait and three of four children also express this trait, the trait is probably due to a dominant allele. In contrast, if both parents do not express a trait and one of their children does, that trait is usually due to a recessive allele. The following pedigree illustrates a sex-linked trait like the color-blindness trait described earlier:

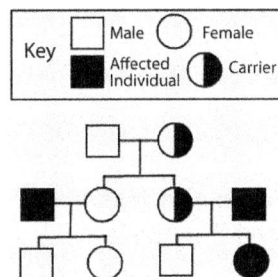

Genetic disorders

Sickle-cell anemia is characterized by weakness, heart failure, joint and muscular impairment, fatigue, abdominal pain and dysfunction, impaired mental function, and death. The mutation that causes this genetic disorder is a point mutation in the sixth amino acid of hemoglobin, which results in an amino acid change from glutamic acid to valine. This mutation causes the chemical properties of hemoglobin to change, exhibiting a sickle shape and having a lower affinity for oxygen.. The sickle shape of the red blood cell makes it difficult to pass through capillaries, and therefore often can clog the blood vessels..

Cystic fibrosis is the most common genetic disorder of people with European ancestry. This disorder affects the exocrine system. A fibrous cyst forms on the pancreas that blocks the pancreatic ducts causing sweat glands to release high levels of salt. A thick mucus, secreted from mucous glands, accumulates in the lungs. The accumulation of mucus causes bacterial infections and possibly death. Scientists have identified a mutation in the protein that transports chloride ions across cell membranes in patients with cystic fibrosis. The majority of the mutant alleles have a deletion of the three nucleotides coding for phenylalanine at position 508. Other people with the disorder have mutant alleles caused by substitution, deletion, and frameshift mutations. Cystic fibrosis cannot be cured, but can be treated for a short time. Most children with cystic fibrosis die before adulthood.

Hemophilia is an inheritable genetic disorder characterized by an inability of the blood to clot. Individuals with hemophilia have lower levels of clotting factors present in their blood. When a blood vessel is damaged, a clot will not form, causing excessive bleeding. This occurs in both external (e.g. skin) and internal (e.g., muscles, joints, or brain) injuries. Hemophilia is a heritable trait because it is passed from mother to child (most commonly male offspring) on the maternal X chromosome. The paternal Y chromosome does not have a gene for this trait. Nevertheless, the paternal X chromosome has the ability to block the trait (dominant) or, by not blocking it, to passively enable it. Hemophilia is an X-linked trait because it is always associated with a deficiency on the X chromosome. As described earlier, women have two X chromosomes. As long as one chromosome is active, she will be unaffected, although she will be a carrier and is capable of passing on the recessive defective gene. Because a male has only one X chromosome, and the Y has no gene for this trait, he is far more likely to be affected by his deficient gene, thus hemophilia is more common in men.

3.A.4: The inheritance pattern of many traits cannot be explained by simple Mendelian genetics.

Non-Mendelian inheritance

Several patterns of genetic inheritance do not conform to Mendel's laws. One example is clonal reproduction. When bacteria reproduce, for example, the offspring are exact genetic copies of the parents. This is also true for mitochondria, the organelles that allow us to harness the energy released by burning organic molecules in oxygen. Mitochondria evolved from symbiotic bacteria, as did the chloroplasts of plant cells, and they reproduce as bacteria do—as genetic copies of a single parent.

Sperm cells are very small, containing little other than the genome they will inject into the egg, and the mitochondria they do have are located in the mid-tail region of the cell, which is lost during fertilization. Epidemiological studies have confirmed that, in humans, the mitochondria received by the embryo, come from the mother. Therefore, whereas your genomic DNA is a mix of your father's and your mother's genes, your mitochondrial DNA is essentially identical to your mother's.

Enduring understanding 3.B: Expression of genetic information involves cellular and molecular mechanisms

3.B.1: Gene regulation results in differential gene expression, leading to cell specialization.

Regulation of genes occur in exons (the protein-coding or functional-RNA-coding parts of genes), **promoters** (the "on" switches for gene expression), **terminators** (the "off" switches for gene expression), or **enhancers** (which are located in other genomic regions and stimulate transcription of genes when bound by transcription factors). There are wide swaths of non-coding DNA that are poorly understood, but have critical functions. Introns tend to be less important; they are spliced out before a mature mRNA is made, and are often nonfunctional, so mutations in introns tend not to have significant effects on organisms in terms of gene regulation. When the mutations are present in exons and important regulatory regions, their effect would depend on the particular gene. Types of mutations and their effects will be discussed in section 3.C.

Many traits are the product of multiple genes and/or physiological processes, and some traits result from nonnuclear inheritance.

DNA regulatory sequences, regulatory genes, and small regulatory RNAs are involved in gene expression.

Regulation of genes can be by positive or negative control. Small molecules can act as activators or inhibitors of gene expression. These regulatory proteins can bind to the DNA to activate transcription and expression; or bind to the DNA and inhibit transcription and expression. These particular proteins are called **transcription factors** because they bind to specific-sequences of DNA and control the rate of transcription as discussed earlier in this chapter.

3.B.2: A variety of intercellular and intracellular signal transmissions mediate gene expression.

Other regulators of gene expression are small proteins called **cytokines**, that are released by cells and affect the activity and function of other cells. Cytokines bind to receptors and are able to induce or repress gene expression. Cytokines play a role in development by allowing cell replication and division. They are also important to immune responses. For example, in the presence of dangerous microbes, cytokines are produced by macrophages in order to activate T cells to fight off infection.

Signals from regulators sometimes require a messenger to amplify the signal. Cyclic AMP is synthesized from ATP and is involved in the regulation of glycogen metabolism. Cyclic AMP relays the signaling cascade of hormones such as glucagon or epinephrine, which results in transforming glucose into glycogen to be stored in the liver. This facilitates the production of free glucose that can be released to the blood.

We previously discussed apoptosis and the possibility that disrupted apoptosis could lead to cancer. A protein called p53 is an important regulator of the cell cycle and is known as a tumor suppressor. When a cell needs to repair DNA, the cell produces increased p53 to fix the DNA, arrest cell growth, or promote apoptosis. Therefore, if p53 is absent or dysfunctional, proliferation of cells with abnormal DNA can proceed, and develop into a cancer.

In embryogenesis, signaling molecules called **morphogens**, diffuse through the tissue to induce cell differentiation and patterning. Gene expression directing a cell towards one fate versus another, is dependent on the concentration of the morphogen as it is distributed throughout the tissue. An example is a morphogen called Sonic Hedgehog, which has been implicated in the proper development of mammalian limbs. Other genes briefly discussed in section 2.E.1 are responsible for patterning and are called **homeobox genes**. They produce DNA-binding proteins responsible for directing cells into particular cell types and structures during development. Researchers have found that mutations in particular homeobox genes called **HOX genes** will result in the development of an alternate body part, like a leg instead of an antenna in a fruit fly.

Enduring understanding 3.C: The processing of genetic information is imperfect and is a source of genetic variation

Inheritable changes in DNA are called **mutations**, which can arise from errors in replication or a spontaneous rearrangement of one or more nucleotide segments. Mutations can also occur from radioactivity, drugs, or chemicals. The severity of the change is not as critical as the location of the change. DNA contains large segments of non-coding areas called introns, but the important coding areas are called exons. If an error occurs in

Signal transmission within and between cells mediates gene expression and cell function.

an intron, it has no effect. If the error occurs in an exon, it has the potential to be lethal, or have no effect, depending on the severity of the mistake. Mutations may occur in the DNA of somatic or sex cells. Usually mutations in sex cells are more dangerous since they contain the blueprint for the developing offspring. Mutations are not always bad. They are the basis of evolution and can be beneficial to the organism if they create a favorable variation that enhances survival. However, mutations may also lead to abnormalities, birth defects, and even death. There are many types of mutations. Let us suppose a normal sequence was as follows:

Normal	A B C D E F	
Duplication	A B **C C** D E F	one nucleotide is repeated
Inversion	A **E D C B** F	a segment of the sequence is flipped
Deletion	A B C E F	a nucleotide is left out (D is lost)
Insertion	A B C **R S** D E F	nucleotides are inserted or translocated
Breakage	A B C	a piece is lost (DEF is lost)

A **point mutation** is a mutation involving a single nucleotide or a few adjacent nucleotides. Deletion and insertion mutations that shift the reading frame are called **frame shift mutations**. A **silent mutation** alters the nucleotide sequence but does not change the amino acid sequence, therefore it does not alter the protein. A **missense mutation** results in an alteration in the amino acid sequence. The effects of the mutation on protein function depend on which amino acids are involved and how many are involved. Since the structure of a protein usually determines its function, a mutation that does not alter the structure will probably have little or no effect on protein function. However, a mutation that does alter the structure of a protein and severely affects protein activity is called a **loss-of-function mutation**. Sickle-cell anemia and cystic fibrosis are examples of loss-of-function mutations.

As discussed, mutations can be lethal, beneficial, neutral, or problematic but not lethal. An example of the last category is the loss of the ability of humans to make their own vitamin C. Most vertebrate species can synthesize vitamin C. Humans also have the genes to do this, but there can be a mutation in the gene encoding the enzyme responsible for the last step in the vitamin C synthesis pathway, L-gulono-|-lactone oxidase. While it would be a health advantage to be able to synthesize vitamin C, it is not a deadly mutation as long as we have access to fresh fruit and vegetables.

3.C.1: Changes in genotype can result in changes in phenotype.

Many types of mutations would be deleterious under some circumstances but beneficial in others. It all depends on the environment and whether or not a trait enhances fitness in that environment. For example, genes that allowed mammals to produce their own body heat, for example, would be wasteful in hot environments, but were critically beneficial in cold, ice-age environments. Many cultivated crops have mutations that would be deleterious in the wild, but are beneficial to humans, so humans cultivate these plants. In the context of this plant-human symbiosis, the agricultural cultivars thrive—even seedless varieties. Modern protozoans and insects have mutations that would never be selected under natural conditions—such as chloroquine resistance in malaria parasites and insecticide resistance in crop pests, fleas, and mosquitoes—but these mutations are critical

Changes in genotype may affect phenotypes that are subject to natural selection. Genetic changes that enhance survival and reproduction can be selected by environmental conditions.

for survival under the circumstances of human use of pesticides.

Non-eukaryotic organisms (bacteria and archaea) have circular chromosomes. Since they reproduce by dividing (producing clones of the parent) the "mixing-up" of genetic information that is so important for evolution must be done in a separate step. Bacteria do this by passing small, modular "mini-chromosomes", known as **plasmids**, to each other. When one bacterium has received a plasmid from another, they will be able to express the genes on that plasmid and therefore will adopt the phenotype encoded by those genes. At this point, the bacterium is said to have been **transformed**. Genetic engineers make use of this property of cells in order to introduce traits of interest. In nature, whole populations of bacteria can rapidly evolve using this mechanism. For example, if you take an antibiotic and one bacterium in a population has a resistance gene, it can pass copies of that gene on to all of its "family members", leading to an antibiotic resistant strain.

3.C.2: Biological systems have multiple processes that increase genetic variation.

The contribution of mutations to variation in a population is minimal. Instead, it is the unique **recombination** of existing alleles that is responsible for the vast majority of genetic differences. Recombination involves crossing over of chromosomes during meiosis resulting in gametes with chromosomes that are unique compared to the parents. DNA change due to recombination, combined with the shuffling of mother's and father's chromosomes during sexual reproduction, results in extensive genetic variation and an array of highly unique offspring.

The main evolutionary advantage of sexually reproducing organisms is the ability to produce offspring that are different from one another, demonstrating **genetic variation** rather than cloning. There are three mechanisms responsible for this: **random assortment of chromosomes** (chromosomes are sorted into daughter cells randomly), **crossing over** (the chromosomes swap portions of DNA); and **random fertilization** (chance alone is responsible for which sperm meets which egg).

3.C.3: Viral replication results in genetic variation, and viral infection can introduce genetic variation into the hosts.

Microbiology includes the study of monera, protists, and viruses. Although viruses are not classified as living things, they greatly affect other living things by disrupting cell activity. **Viruses** are obligate parasites because they rely on the host for their own reproduction. Viruses are composed of a protein coat and a nucleic acid, either DNA or RNA. A bacteriophage is a virus that infects a bacterium. Animal viruses are classified by the type of nucleic acid, presence of RNA replicase, and presence of a protein coat.

There are two types of viral reproductive cycles:

1. **Lytic cycle** — The virus enters the host cell and makes copies of its nucleic acid and protein coat, and then reassembles. Afterward, it lyses or breaks out of the host cell and infects other nearby cells, repeating the process.
2. **Lysogenic cycle** — The virus may remain dormant within the cell until specific factors activate and stimulate it to break out of the cell. Herpes is an example of a lysogenic virus.

The imperfect nature of DNA replication and repair, along with horizontal acquisitions of genetic information, increases variation.

Sexual reproduction in eukaryotes involving gamete formation, including crossing-over during meiosis and the random assortment of chromosomes during meiosis, and fertilization serve to increase variation. Reproduction processes that increase genetic variation are evolutionarily conserved and are shared by various organisms.

A special type of virus called a **retrovirus** uses RNA to make DNA that can integrate into the host genome. HIV is a type of retrovirus, and since retroviruses are easily mutated, it is one reason that it is difficult to generate a treatment to eradicate HIV infection.

Enduring understanding 3.D: Cells communicate by generating, transmitting and receiving chemical signals ____

3.D.1: Cell communication processes share common features that reflect a shared evolutionary history.

Hormones regulate sexual maturation in humans. Humans cannot reproduce until puberty, about the age of 10–14, depending on the individual. The hypothalamus begins secreting hormones that stimulate maturation of the reproductive system and development of secondary sex characteristics. Reproductive maturity in girls occurs with their first menstruation and occurs in boys with the first ejaculation of viable sperm.

Hormones also regulate reproduction. In males, the primary sex hormones are the androgens, testosterone being the most important. The androgens are produced in the testes and are responsible for primary and secondary sex characteristics of the male. Female hormone patterns are cyclic and complex. Most women have a reproductive cycle length of about 28 days. The menstrual cycle is specific to the changes in the uterus. The ovarian cycle results in ovulation and occurs in parallel with the menstrual cycle. This parallelism is regulated by hormones. Five hormones participate in this regulation, most notably estrogen and progesterone. Estrogen and progesterone play an important role in signaling to the uterus and development and maintenance of the endometrium. Estrogens are also responsible for secondary sex characteristics of females.

3.D.2: Cells communicate with each other through a direct contact with other cells or from a distance via chemical signaling.

Signals may come from the environment, or they may come from other cells. Junctions that connect the cytosol of adjacent cells and allow the transfer of small molecules and ions are called **plasmodesmata** in plants, and **gap junctions** in animals. Furthermore, two cells can also communicate with each other through molecules expressed on their surfaces. This is the case for several types of immune cells, including T-cells.

Hormones are another type of signal that serve to relate messages between different types of cells—even in far off regions of the body! The **endocrine system** is a collection of glands that produce hormones that are released into the bloodstream and are carried to a target tissue where they stimulate an action. There are two classes of hormones: steroid and peptide. Steroid hormones are derived from cholesterol and include the sex hormones. Peptide hormones are derived from amino acids. Hormones have a specific function, and fit receptors on the cell surface in the target tissue. Receptor binding then activates an enzyme that converts ATP to cyclic AMP. Cyclic AMP (cAMP) is a second messenger that travels from the cell membrane to the nucleus. When cAMP is triggered, the genes found in the nucleus turn on or off to cause a specific response.

Hormones are secreted by cells found in endocrine glands. The major endocrine glands and their hormones are described on the following page:

Viral replication differs from other reproductive strategies and generates genetic variation via various mechanisms.

Correct and appropriate signal transduction processes are generally under strong selective pressure.

Signals released by one cell type can travel long distances to target cells of another cell type.

Hypothalamus – located in the lower brain. Hypothalamus produces a number of important hormones including antidiuretic hormone (ADH) and oxytocin. These are released by the posterior lobe of the pituitary gland and function to promote water retention by the kidneys and stimulate contractions in the uterus and mammary glands, respectively. The hypothalamus also regulates the anterior pituitary gland to signal the production of more hormones that you will read about below.

Pituitary gland – located at the base of the hypothalamus and divided into two lobes–posterior and anterior. It is regulated by the hypothalamus and releases hormones including growth hormone that stimulates bone growth and metabolic function; prolactin, which stimulates milk production by the mammary glands; and follicle-stimulating hormone (FSH) and lutenizing hormone (LH) which are both important for regulation of the reproductive organs.

Thyroid gland – located in front of the trachea; The thyroid gland produces hormones, including thyroxine, that help maintain heart rate, blood pressure, muscle tone, digestion, and reproductive functions. It also releases calcitonin, which functions to lower blood calcium levels. The parathyroid gland, located just above the thyroid, releases parathyroid hormone to raise blood calcium levels.

Adrenal gland – one located above each kidney; The adrenal glands release hormones in response to stress or anxiety, including epinephrine and norepinephrine. They stimulate increases to your heart rate, blood pressure, and blood glucose level, as well as constricting some blood vessels. You may have heard epinephrine called adrenaline—don't let your adrenal glands take over when you are taking the AP Biology Exam!

Pancreas – located behind the stomach and secretes insulin to lower blood glucose levels. It also secretes glucagon to raise blood glucose levels.

Gonads – the ovaries of the female and the testes of the male. The three gonadal steroids, androgen (testosterone), estrogen, and progesterone, regulate the development of the male and female reproductive organs. Ovaries release estrogens to stimulate uterine lining growth and promote female characteristics. The ovaries also release progesterone to promote uterine lining growth. Testes

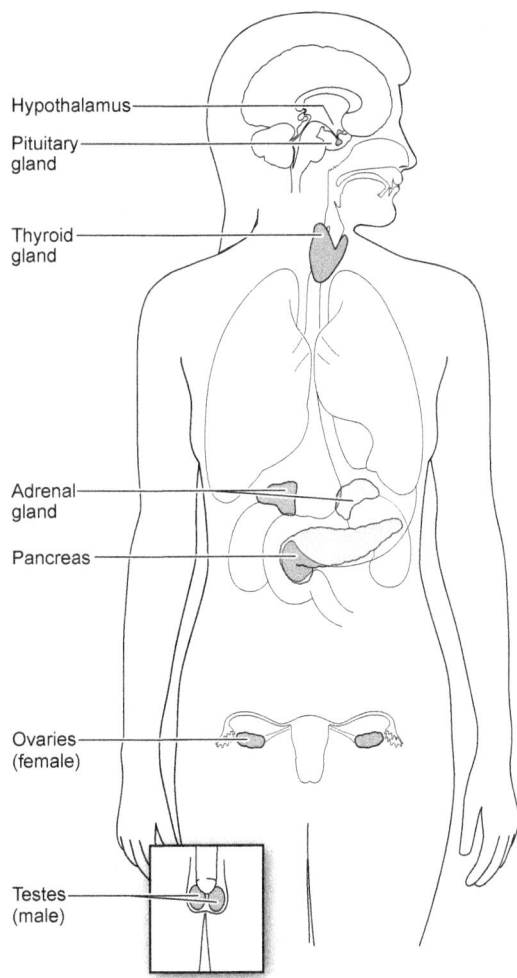

Hypothalamus

Pituitary gland

Thyroid gland

Adrenal gland

Pancreas

Ovaries (female)

Testes (male)

release androgens to support sperm formation and to promote male characteristics.

In addition, some neurotransmitters (like acetylcholine), can act as hormones. Acetylcholine controls muscle contraction and heartbeat resulting in profound effects on the cardiovascular and respiratory systems. These hormones/neurotransmitters can be used to increase the rate and stroke volume of the heart, thus increasing the rate of oxygen delivery to the blood cells.

It is important to note that these signals can travel long distances. For example, glucose regulation can be achieved by circulating factors in the blood that signal to the cells responsible for adjusting glucose levels by secreting insulin.

3.D.3: Signal transduction pathways link signal reception with cellular response.

Cells communicate by sending and receiving signals and have evolved a variety of signaling mechanisms to accomplish the transmission of important biological information. Evolutionarily, sending the proper signals is strongly selected by natural selection. The transmission of signals by molecules from outside of the cell to the inside is initiated by cell-surface receptors, and is a process called **signal transduction**.

Signaling in multicellular organisms helps to coordinate individual cells to allow the whole organism to function properly. Signal transduction begins when the chemical signal called a **ligand**, is recognized by a receptor protein. The chemical signal can be a peptide or a small protein. The receptor protein is often the initiation point for a signaling cascade that ultimately results in a change in gene expression, protein activity, or physiological state of the cell or organism, including cell death (apoptosis). Once the ligand binds to its receptor, the shape of the protein changes, allowing the signal to be converted to a response by the cell. Three major types of receptors exist:

- **G-protein linked receptors** bind the energy-rich protein Guanosine-5'-triphosphate (GTP).
- **Ligand-gated ion channels** play an important role in the nervous system, as neurotransmitters released at a synapse allow the channel to open and propagate the signal.
- **Receptor tyrosine kinase**s are cell membrane receptors that exhibit enzymatic activity through kinases that can phosphorylate proteins, resulting in their activation and a signaling cascade.

Secondary messengers are intracellular signaling molecules released by the cell to trigger physiological changes such as proliferation, differentiation, migration, survival, and apoptosis. Secondary messengers are therefore one of the initiating components of intracellular signal transduction cascades. Examples of secondary messenger molecules include cyclic AMP, cyclic GMP, inositol trisphosphate (IP3), diacylglycerol, and calcium.

3.D.4: Changes in signal transduction pathways can alter cellular response.

Defects in any part of the signaling pathway through genetic changes can lead to severe or detrimental conditions such as impaired embryonic development, autoimmune diseases, or cancer. Poisons or neurotoxins, as well as anesthetics, can also change signal transduction efficiency. It has been of great research interest to devise a targeted therapy

Signal transduction is the process by which a signal is converted to a cellular response.

for cancer in which molecules can be used to impact known factors or cellular signaling pathways that are involved in the progression of cancer. For example. it is thought that if one could limit the development of the blood supply within a tumor, then the tumor cells would not receive the nutrients required to survive and therefore the tumor should shrink. In this case, site-specific targeting and inhibition of molecules that are critical to blood vessel development may be a useful therapy for cancer. It should be noted, that this type of targeting is not easy—some molecules could be part of multiple or complex signaling pathways, and this would add to the complications of determining a new treatment for cancer. Taken together, although there are some promising anti-angiogenic molecules, this is an area that requires further research!

Enduring understanding 3.E: Transmission of information results in changes within and between biological systems

3.E.1: Individuals can act on information and communicate it to others.

Communication is as important as breathing or feeding. To survive in nature and increase fitness, all organisms must be able to "read" their environment and to extract information about food, shelter, mating partners, and threats, as well as changes in the abiotic properties of the environment. Species that live in colonies or herds usually employ visual or acoustic behavior to warn the other members that a predator is nearby. Animals may also release stress pheromones, which alerts conspecifics in the vicinities that they should flee. The immediate response of an animal under attack by a predator is the fight-or-flight response. It involves a cascade of hormone release, controlled by the sympathetic nervous system, that enables a quick and active response. Another way of deceiving predators is by playing dead, through a behavior called **tonic immobility**. This strategy works when the predator loses interest in an immobile, potentially dead prey. Possums, grasshoppers, and woodlice are examples of animals that use this strategy.

Communication is also used to establish and maintain a territory. Mammals are good examples of territorial animals. The advantage of being able to establish a territory, or to recognize someone else's territory, is to lessen the probability of encountering a competitor, which could lead to fights (i.e. waste of energy, injuries), and even death.

Communication is not only used to avoid predators or competitors, it also plays an important role in cooperative behavior. An individual bee uses a series of movements called a "waggle dance" to inform the other members of its colony where it found nectar, water, or potential new nesting sites. Using the dance, bees pass information about the direction and distance to the resource. Ants perform a very similar action, but they rely on chemical cues. By laying down a trail of pheromones, scout ants inform the nestmates about the location of a new resource.

Courtship and mating behavior are strongly built upon diverse communication cues, which are used to determine the presence, or the quality of a partner. Sex pheromones are widespread in nature, and help to locate a sexually mature partner. Courtship displays are usually employed by males to conquer the right to copulate with a female. Females, in turn, choose the males by assessing direct and indirect benefits. A **direct benefit** is an overtly observed trait that indicates a direct benefit on survival in the process of choosing

Conditions where signal transduction is blocked or defective can be deleterious, preventative or prophylactic.

Responses to information and communication of information are vital to natural selection and evolution.

a mate. An example is strong and active males that may be more efficient in foraging and protecting against predators. **Indirect benefits** are those that will increase the offspring fitness, such as genetic information coding for favorable traits. Sexual dimorphism often arises when visual clues are employed by the organism. For instance, male fiddler crabs always have one claw that is disproportionally longer than the other. They use this claw to wave at the female during sexual display. The waving movements also inform the female about the health and strength of the male, and therefore females tend to chose males with larger claws, as they will have an advantage in building the burrows needed for reproduction.. One of the most remarkable courtship behaviors is exhibited by birds-of-paradise. The male birds employ a combination of visual cues (plumage), songs, and ritualized movements to attract a potential partner. The females assess all of these clues to decide whether they should mate with the male.

3.E.2: Animals have nervous systems that detect external and internal signals, transmit and integrate information, and produce responses.

The **central nervous system (CNS)** consists of the brain and spinal cord. The CNS is responsible for the body's response to environmental stimuli. The spinal cord sends out motor commands in response to stimuli that are automatic or by reflex. Responses to more complex stimuli occur in the brain. The meninges are the connective tissues that protect the CNS. The CNS contains fluid filled spaces called ventricles that are filled with cerebrospinal fluid which is formed in the brain. Cerebrospinal fluid cushions the brain and circulates nutrients, white blood cells, and hormones.

The **peripheral nervous system (PNS)** consists of the nerves that connect the CNS to the rest of the body. It has two divisions: sensory, which brings information to the CNS from sensory receptors; and motor, which sends signals from the CNS to effector cells. The motor division is further divided into the somatic nervous system and the autonomic nervous system. The somatic nervous system is controlled consciously in response to external stimuli. The autonomic nervous system regulates the body's internal environment by unconsciously controling the hypothalamus of the brain. This system is responsible for the movement of the heart and other smooth muscle organs and organ systems.

The **neuron** is the basic unit of the nervous system. It consists of an axon, which carries impulses away from the cell body; the dendrite, which carries impulses toward the cell body; and the cell body, which contains the nucleus. The myelin sheath, comprised of Schwann cells, covers the neuron and provides insulation, which allow electrical impulses to travel quickly through the neuron. Synapses are the junctions between neurons.

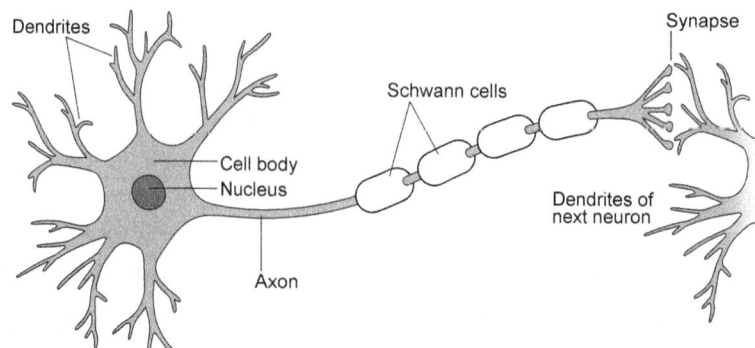

Chemicals called neurotransmitters serve as signaling molecules that are released from one neuron and diffuse through the synaptic cleft to send signals to another neuron. The most common neurotransmitter is acetylcholine. Acetylcholine controls muscle contraction and heartbeat. As we discussed in section 3.D.2, norpinephrine and epinephrine are catacholemines—neurotransmitters produced in response to stress. The major inhibitory neurotransmitter in the mammalian central nervous system is Gamma-Aminobutyric acid (GABA) which reduces neuronal excitability throughout the nervous system and has profound effects on the cardiovascular and respiratory systems. These neurotransmitters are used to increase the rate and stroke volume of the heart, thus increasing the rate of oxygen delivery to the blood cells. Furthermore, the nervous system can release another neurotransmitter, dopamine, which plays a major role in reward-motivated behavior.

Nerve action depends on an imbalance of electrical charges between the inside of the neuron and the outside. When a neuron is resting, it has a negative charge and is said to be hyperpolarized. When ions like sodium, potassium, or calcium flow into the neuron, it takes on a positive charge and becomes depolarized. When the ions move from one side of the neuronal membrane to the other (from outside the cell to inside, or vice versa) an electrical current flows through the neuron. These electrical currents are called action potentials. **Action potentials** trigger the release of neurotransmitters from the axon into the synaptic cleft. When the neurotransmitters diffuse through the synaptic cleft, they bind to receptors on the surface of dendrites. This binding then triggers another action potential in the next neuron.

Whereas action potentials propagate impulses along neurons, transmission of information between neurons occurs across synapses.

Some neurons, or nerves, form synapses on the muscle cells. This is called a neuromuscular junction. In neuromuscular junctions there is a threshold of neurotransmitters that must be released by the nerve in order to generate a response from the muscle cell. This is called an "all or none" response.

Disorders of the nervous system have a variety of causes, and have effects across many physiological systems. One such disorder is Parkinson's disease, in which there is a gradual loss of neurons, largely resulting in effects on movement. The transmission of motor impulses to the muscles is negatively impacted by the loss of neurons that produce dopamine. Symptoms include tremors, slow movement, and muscle rigidity. While there is no cure for Parkinson's disease, medications are available that increase levels of dopamine. These are dopamine agonists—meaning that they mimic the effects of dopamine to stimulate brain function and control of motor movements. Deep brain stimulation is a surgical treatment that has been observed to reduce symptons of Parkinson's disease. Due to side effects and the build-up of tolerance to some of the medications, researchers continue to search for better treatments for the disease. Among the many current research projects focusing on Parkinson's disease, studies on stem cells have been designed to target the dying neurons, and generate new neurons to produce dopamine, thereby limiting the effects of Parkinson's disease. Promising results have been observed with stem cell therapy, however further research is needed.

Keywords

Genotype
Phenotype
Nucleic acids
  DNA (deoxyribonucleic acid)
  RNA (ribonucleic acid)
DNA replication
  RNA polymerase
  Leading strand
  Lagging strand
  Helicase
  Replication fork
  RNA polymerase
  Okazaki fragments
  DNA ligase
Plasmid
  Conjugation
Messenger RNA
Ribosomal RNA
Transfer RNA
Transcription
  Promoter
  Sense
  Antisense
  Initiation
  Elongation
  5' capping
  3' polyadenylation
  Intron removal
Translation
  Genetic code
  Codons
  Anticodons
Genetic engineering
Polymerase Chain Reaction (PCR)
Gel electrophoresis
Cell Cycle
  Interphase
    G1 phase
    S phase
    G2 phase
  Haploid
  Diploid

Mitosis
  Prophase
  Metaphase
  Anaphase
  Telophase
  Cytokinesis
Meiosis
  Prophase I/II
  Metaphase I/II
  Anaphase I/II
  Telophase I/II
Law of Segregation
  Homozygous,
  Heterozygous
  Punnett square
Law of Independent Assortment
Inheritance
  Monohybrid
  Dihybrid
  Sex-linked
Promoter
Terminator
Enhancer
Transcription Factor
Recombination
  Random assortment of
    chromosomes
  Crossing over
  Random fertilization
Mutations
  Duplication
  Inversion
  Deletion
  Insertion
  Breakage
  Silent mutation
  Missense mutation
Virus
  Lytic cycle
  Lysogenic cycle
Retrovirus
Plasmodesmata

Gap junctions
Endocrine system
   Hypothalamus
   Pituitary gland
   Thyroid
   Pancreas
   Gonads
Signal transduction
   Ligands
   G-protein linked receptors
   Ligand-gated ion channels

Receptor tyrosine kinases
Gene expression
Cytokines
Tonic immobility
Direct/Indirect benefits
Nervous System
CNS/PNS
Somatic/Autonomic
Neurons
Neurotransmitter
Action potential

Summary

Information that drives the continuity of life is passed on through generations via DNA, and changes in DNA sequences can lead to heritable mutations.

Structure and function of cells and organisms require signals and cell signaling mechanisms to regulate gene expression and maintain the proper health of the organism. Genetic engineering provides potential to correct these heritable changes.

Mitosis, and the production of identical daughter cells, ensures that heritable information is maintained. Meiosis, and the production of reproductive cells, ensures that information from both parents is recombined after fertilization – contributing to natural selection, and the unique phenotypes observed in offspring.

Mathematical formulas described by Mendel can be used to describe the inheritance of traits, but phenotypes are not always expressed in Mendelian frequency.

Sensory organs have evolved, including those within the endocrine and nervous systems, to detect external information and coordinate transmission and processing of the signals.

Location and function of important hormones

Location	Hormone	Function
Hypothalamus	Antidiruetic hormone	Promotes water retention by kidneys
	Oxytocin	Stimulates mammary gland cells and uterus contractions
Pituitary gland	Growth hormone	Stimulates growth during childhood
	Prolactin	Associated with milk production by mammary glands
	Follicle-stimulating hormone	Stimulates maturation of eggs in ovary and lining of the uterus
	Lutenizing hormone	
Thyroid gland	Calcitonin	Lowers blood calcium level
	Thyroxine	Maintains metabolic processes
Parathyroid	Parathyroid hormone	Raises blood calcium level
Adrenal gland	Epinephrine	Increases metabolic rate and glucose levels, and constricts blood vessels
	Norepinephrine	
Pancreas	Insulin	Lowers blood glucose levels
Ovaries (female)	Estrogen	Maintain female reproductive organs
	Progesterone	Maintain female reproductive organs
Testes (male)	Androgens (Testosterone)	Maintain male reproductive organs

1. Over a 20-year span of time, a female whale produced all male offspring. She is pregnant again. What is the probability that this next offspring will be male?

   A. 10%

   B. 50%

   C. 75%

   D. 100%

Questions 2 and 3 refer to the following information:

Achondroplasia is a dominant genetic trait that causes dwarfism (stunted growth). The homozygous condition is lethal. Heterozygotes, however, express the dwarf trait. Tristan has a family history of dwarfism, but does not express the dwarf trait himself. His wife, Penelope, has achondroplasia.

2. What are the respective genotypes of Tristan and Penelope?

   A. Aa, aa

   B. aa, Aa

   C. Aa, AA

   D. aa, AA

3. What is the probability their child will have achondroplasia?

   A. 0 %

   B. 25%

   C. 50%

   D. 75%

4. If one parent has the genotype AABBCCDDEE and the other parent has the genotype aabbccddee, what are the possible genotypes of their offspring?

   A. All AaBbCcDdEe

   B. AABBccDDEE and aaBBCCDDEE

   C. aaBBccDDee and AAbbCCddEE

   D. AaBBCcDDEE and aABbCCDdEE

5. **An organism was found to have DNA that contained 20% cytosine. Which of the following can be concluded about the DNA of this organism?**

   A. It has 30% guanine.

   B. It has 80% guanine.

   C. It has 40% adenine and 40% thymine.

   D. It has 30% thymine and 30% adenine.

6. **Gregor Mendel was an Austrian monk who studied the inheritance patterns of different characteristics of peas. He discovered that green coloring (G) is dominant to yellow (g) and that round seeds (R) are dominant to wrinkled (r). What would be the projected phenotypic ratio for a cross between two parents that were heterozygous for both traits?**

   A. All peas would be green and round.

   B. Half of the peas would be green/round, and half would be yellow/wrinkled.

   C. There would be 1 green/round, 3 green/wrinkled, 3 yellow/round, 9 yellow/wrinkled.

   D. There would be 9 green/round, 3 green/wrinkled, 3 yellow/round, 1 yellow/wrinkled.

7. **What would the complementary strand be to a piece of DNA with the nucleotide sequence AGGTCCGATCA?**

   A. AGGTCCGTCA

   B. GAACTTAGCTG

   C. GTTCGTAACGT

   D. TCCAGGCTAGA

8. **When studying the genetics of a bacterium, a scientist found its DNA sequence to be AGTTCGCTATCCA. After irradiating the bacterium, the DNA sequence was AGTTCTATCCA. What type of mutation has occurred?**

   A. Frame-shift mutation

   B. Nonsense mutation

   C. Deletion mutation

   D. Missense mutation

9. The DNA sequence of a cell is GCCGTATAGCA. What would be the corresponding strand of mRNA to attach to this strand during transcription?

   A. CGGCATATCGT

   B. CGGCAUAUCGU

   C. AUUACGCGAUC

   D. TCCTCGCGATA

Questions 10-11 refer to the following information:

Humans can have either attached earlobes (f) or free earlobes (F). Assume that two parents are heterozygous for free earlobes.

10. What are the genotypes of the parents?

   A. Ff × Ff

   B. Ff × ff

   C. ff × Ff

   D. FF × ff

11. Which statement about the possible phenotypes of the offspring is correct?

   A. All offspring will have free earlobes.

   B. All offspring will have attached earlobes.

   C. 25% of the offspring will have free earlobes.

   D. 75% of the offspring will have free earlobes.

Chapter 3 Quiz Answer Key _____

1. **Answer: B.** Since whales have two possible genders, there is always a 50/50 chance of male versus female. Having several male offspring in a row does not increase the chances of producing a female calf.

2. **Answer: B.** To not show achondroplasia and still be alive, Tristan needs to be homozygous recessive (aa). Penelope exhibits stunted growth, so she has to be heterozygous (Aa).

3. **Answer: C.** Tristan cannot pass the gene to his offspring because he does not have one. Penelope has a 50% chance of passing either the dominant or the recessive allele. Therefore, the genotype of their offspring would either be aa (no dwarfism) or Aa (dwarfism).

4. **Answer: A.** Each parent is homozygous for all of the traits. This means that the first parent can only pass a dominant allele, and the second parent can only pass a recessive allele to the offspring. Therefore, the genotype of all of the offspring has to be heterozygous for all traits.

5. **Answer: D.** If an organism has 20% cytosine, then it also has to have 20% guanine. This means that the other two nitrogen bases must be equally divided among the other 60%, so 30% for thymine and 30% for adenine.

6. **Answer: D.** When doing a cross of this type it is important to separate the alleles properly so that each has the opportunity to be inherited. In this case, the alleles in the gametes would be GR, Gr, gR, gr. The Punnett square will have 16 possible offspring. All phenotypes will be represented in the ratio of 9:3:3:1.

7. **Answer: D.** A pairs with T, and C pairs with G. Therefore, the complementary strand of the DNA molecule contains the matching pair on the other side.

8. **Answer: C.** The new sequence of DNA is missing two of the nitrogen bases that were present in the original strand. This means the radiation caused a deletion.

9. **Answer: B.** During transcription, the RNA strand matches up with the DNA strand to form a complement. However, in RNA there is no T. It is replaced with U. Therefore, every time A is present in DNA, the RNA matches up with a U.

10. **Answer: A.** Both parents have a dominant and a recessive allele. This is what makes them heterozygous.

11. **Answer: D.** Performing a Punnett square analysis, you can see that there is a 25% chance of an FF genotype, a 50% chance of an Ff genotype, and a 25% chance of an ff genotype. Since only one dominant gene is needed to have free earlobes, those genotypes with an F are free lobes. Therefore, the offspring have a 75% chance to have free earlobes.

Chapter 4.
Big Idea 4: Systems

Big Idea 4: Systems: Biological systems interact, and these systems and their interactions possess complex properties.

What you will learn from this chapter:

- Differences in chemical bonds that affect characteristics of molecules
- Carbohydrates, lipids, and proteins
- Specialized functions of subcellular components
- How species coexist with their resources
- How species diversity affects the ecosystem

Introduction _____

This chapter will focus on both the chemistry and biology of cellular processes. By understanding the key molecules involved and how their structure affects their function, we can better understand human physiology and its dysfunction. Organization at a cellular level – and the components that contribute to the function of a cell, draws parallels to the organization of the trophic levels of the environment and how the various organisms interact with one another. Let's learn what makes the world work together!

Enduring understanding 4.A: Interactions within biological systems lead to complex properties _____

4.A.1: The subcomponents of biological molecules and their sequence determine the properties of that molecule.

Understanding the metabolism of cells is not only critical to understanding how cells function individually, but also how they function together to form a complex system such as a nerve, and even more complex, like the whole nervous system. For this skill set, you will need to start with the basic understanding of how chemicals are formed, and what functions and properties they have once they are combined. By now, you have certainly seen a periodic table containing all of the elements…so let's start there!

An **element** is a substance that cannot be broken down into other substances. An **atom** is the smallest particle of an element that exhibits the properties of the element. All of the atoms of a particular element are the same. The atoms of each element are different from the atoms of the other elements. The core of the atom is called the nucleus and contains particles called protons and neutrons. Protons are positively charged, and neutrons are uncharged. Another, even smaller component of the atom is made up of particles called electrons. Electrons are negatively charged and spin around the nucleus of the atom in orbits or shells. The outermost shell of an atom is most stable when it contains eight electrons. If the outer shell does not have eight electrons, an atom will form bonds with other atoms that can donate or accept electrons to form a bond.

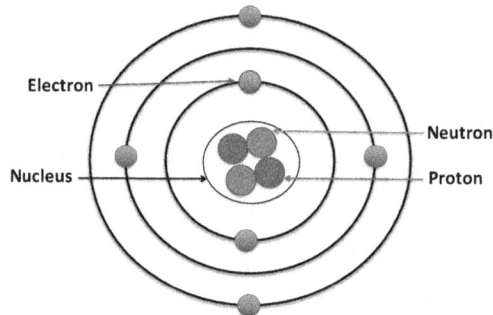

Molecules and Chemical Bonds

A **molecule** is the smallest particle of a substance that can exist independently and has all of the properties of the substance. A molecule of most elements is made up of one atom, and a **compound** is formed when two or more atoms are chemically combined.

Oxygen, hydrogen, nitrogen, and chlorine molecules are made of two atoms each.

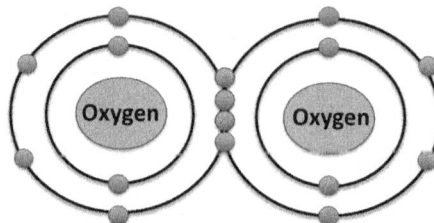

Sometimes the formation of a compound results in different properties—the elements can lose their individual identities. For example, the formation of water by hydrogen and oxygen forms a liquid. The **chemical formula** shows the elements in a compound by using symbols and subscripts. For example, carbon dioxide is made up of one atom of carbon (C) and two atoms of oxygen (O_2), so the formula is CO_2.

Compounds are held together by chemical bonds. They form when atoms with incomplete shells share or completely transfer their electrons to other atoms. There are three types of chemical bonds; hydrogen bonds, covalent bonds, and ionic bonds.

A **hydrogen bond** is the weakest type of bond, and forms when one electronegative atom shares a hydrogen atom with another electronegative (discussed below) atom. An example of a hydrogen bond is a water molecule (H_2O) bonding with an ammonia molecule (NH_3). The H^+ atom of the water molecule weakly attracts the negatively charged nitrogen. Weak hydrogen bonds are helpful to the formation of new bonds because they are easily broken apart due to the brief nature of the bond.

Covalent bonding is the sharing of electrons by two atoms, and is the strongest chemical bond. A simple example of a covalent bond is the bond formed between two hydrogen atoms. Each hydrogen atom has one electron in its outer shell; therefore the two hydrogen atoms come together to share their electrons. Some atoms share two pairs of electrons, like two oxygen atoms resulting in a double covalent bond.

When an atom transfers one or more electrons to another atom, this creates an **ionic bond**. The loss of an electron from donating it to the other atom results in a positively charged atom; and as a result of accepting the donated electron, the other atom becomes negatively charged. These charged atoms are called ions. An example of an ionic bond is sodium chloride (NaCl). A single electron on the outer shell of sodium joins the chloride atom with seven electrons in its outer shell. The sodium now has a +1 charge and the chloride now has a –1 charge. The charges attract each other to form an ionic bond. Ionic compounds are called salts. In a dry salt crystal, the bond is so strong it requires a great deal of strength to break it apart. However, if the salt crystal is placed in water, the bond will dissolve easily as the attraction between the two atoms decreases.

Polarity

When an atom is part of a molecule, it can exhibit features of **electronegativity**, which is the ability of an atom to attract electrons. The greater the electronegativity of an atom, the greater its capability of pulling shared electrons toward itself. Electronegativity of the atoms determines whether a bond is polar or nonpolar. In **nonpolar covalent bonds**, the electrons are shared equally, thus the electronegativity of the two atoms is the same. This type of bonding usually occurs between two of the same atoms. A **polar covalent bond** forms when different atoms join together, for example, the creation of water by hydrogen and oxygen. In this case, oxygen is more electronegative than hydrogen, so the oxygen exerts a stronger pull on the hydrogen electrons.

The four major chemical compounds found in the cells and bodies of living things are: carbohydrates, lipids, proteins, and nucleic acids. Each of these molecules has a carbon skeleton and therefore is an organic compound, whereas molecules without carbon are inorganic.

Monomers are the simplest unit of structure. Monomers combine together to form polymers, or long chains, making a large variety of molecules. Monomers combine through the process of **condensation** reactions (also called dehydration synthesis). In this process, one molecule of water is removed between each of the adjoining molecules.

Structure and function of polymers are derived from the way their monomers are assembled.

In order to break the molecules apart in a polymer, water molecules are added between monomers, thus breaking the bonds between them. This process is called **hydrolysis**.

Carbohydrates are made up of carbon, hydrogen and oxygen always in the ratio of 1:2:1. Therefore, the simplified formula to denote carbohydrate content within a molecule is: $(CH_2O)_n$.

Simple carbohydrates are called **monosaccharides**; saccharide is a word of Greek origin meaning "sugar". The most abundant monosaccharide is glucose, which is produced by plants after converting sunlight to energy and broken down by cells through the process of respiration to release energy. Another type of monosaccharide is fructose, commonly found in fruits.

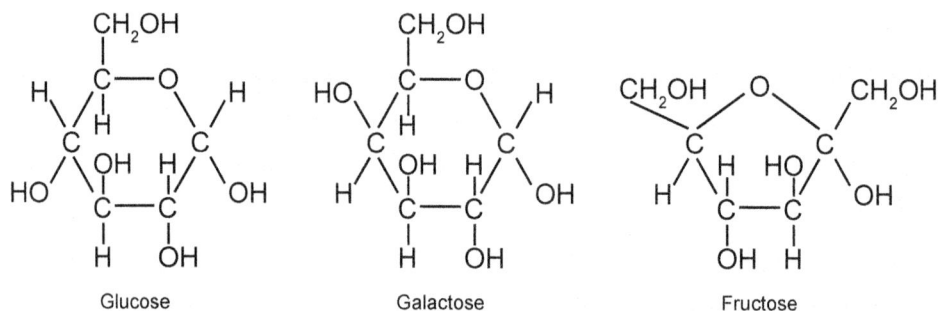

Glucose   Galactose   Fructose

Disaccharides are made by joining two monosaccharides by removal of a water molecule (condensation) to form a chemical link called a glycosidic bond. When disaccharides are broken down to two individual monosaccharides, water is used to break the glycosidic bond through hydrolysis. Maltose is the combination of two glucose molecules, lactose is the combination of glucose and galactose, and sucrose is the combination of glucose and fructose.

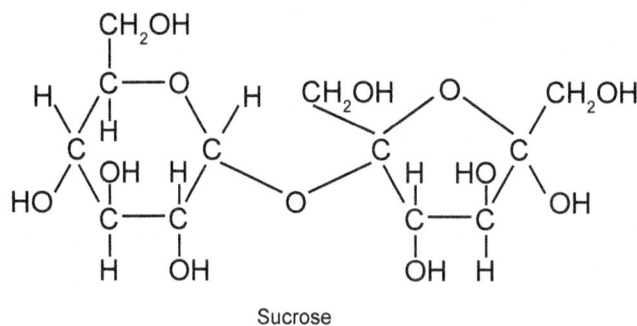

Sucrose

Complex carbohydrates are called **polysaccharides** because they consist of many monomers joined together. They can provide structure for the cell or be stored as energy. Types of polysaccharides include.

Starch: a major energy storage molecule in the plastids of plants

Glycogen: a major energy storage molecule in the liver and muscle cells of animals

Cellulose: a molecule present in plant cell wall to provide structure. Many animals lack the enzymes necessary to hydrolyze cellulose, so it simply adds bulk (fiber) to the diet

Chitin: a molecule found in the exoskeleton of arthropods and fungi

Lipids play an important role in cell membrane structure, energy storage, and insulation. Lipids are **hydrophobic** molecules ("water fearing") and will not mix with water. There are three important families of lipids: fats, phospholipids and steroids.

Fats consist of glycerol and three fatty acids, and are also called "triglycerides." Fatty acids are a carboxylic acid with a long carbon chain. The nonpolar carbon-hydrogen bonds in the tails of fatty acids are highly hydrophobic. Fats cushion and insulate the body and nerves and are solids at room temperature.

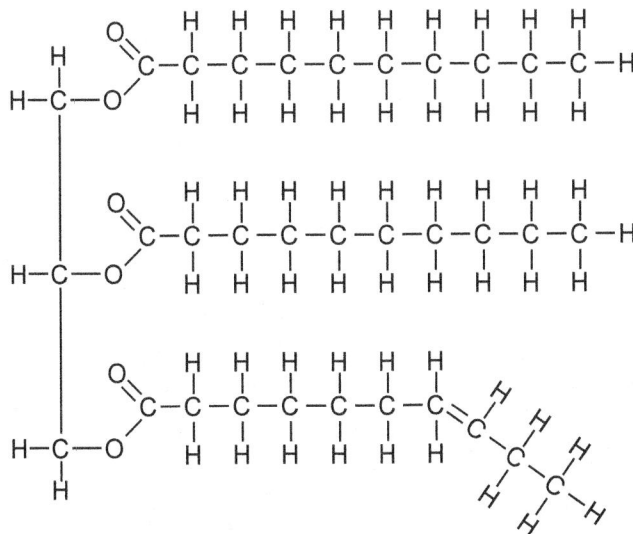

Structure of a fat

Phospholipids are a vital component of cell membranes. In a phospholipid, one or two fatty acids are replaced by a phosphate group linked to a nitrogen group. They consist of a **polar** (charged) head that is **hydrophilic** ("water loving") and a **nonpolar** (uncharged) tail, which is hydrophobic. This allows the membrane to orient itself with the polar heads facing the interstitial fluid found outside the cell and the nonpolar tails facing the internal fluid of the cell. A molecule that contains portions that are both hydrophilic and hydrophobic is called **amphipathic.** Therefore, a phospholipid is amphipathic.

Steroids are composed of a carbon skeleton consisting of four inter-connected carbon rings, and are not soluble in water. An important steroid is cholesterol, which is the precursor from which other steroids are synthesized. Some hormones, including testosterone and estrogen, are steroids.

Proteins are critical molecules of all living cells. They are made up of any combination of twenty **amino acids**, which are joined together by condensation reactions that remove water. The bond formed between two amino acids is called a peptide bond.

An amino acid is made up of four parts: one hydrogen atom, an amino group ($-NH_2$), a carboxyl group ($-COOH$), and a side chain. The amino acid is defined by its side chain, which is also called the functional, or R, group. A table of common functional groups and their corresponding molecular formula is found on the next page.

Functional Group	Molecular Formula
Amino	$-NH_2$
Alkyl	$-C_nH_{2n+1}$
Methyl	$-CH_3$
Ethyl	$-C_2H_5$
Propyl	$-C_3H_7$
Carboxyl	-COOH
Hydroxyl	-OH
Aldehyde	-CHO
Keto	-CO
Sulfhydryl	-SH
Phenyl	$-C_6H_5$
Phosphate	$-PO_4$

Directionality influences structure and function of the polymer.

An analogy can be drawn between the twenty amino acids and the alphabet. We can form millions of words using an alphabet of only twenty-six letters. Similarly, organisms can create many different proteins using the twenty amino acids. Polymers of amino acids are called polypeptide chains. This results in the formation of many different proteins whose structure typically defines their function. Examples of protein function include

- structure and support (e.g., connective tissue, hair, feathers, and quills)
- storage of amino acids (e.g., albumin in eggs and casein in milk)
- transport of substances (e.g., hemoglobin)
- coordination of body activities (e.g., insulin)
- signal transduction (e.g., membrane receptor proteins) (discussed in section 3.D.3)
- contraction (e.g., muscles, cilia, and flagella)
- body defense (e.g., antibodies)
- enzymes to speed up chemical reactions

Proteins can be folded into four types of structures.

The unique linear sequence of amino acids makes up the **primary structure** of the protein. Even the smallest change in primary structure can affect the structure and function of the protein.

Secondary structure refers to the coils and folds resulting from the hydrogen bonds along the polypeptide chains. The secondary structure can take the form of an alpha helix—a coil held together by hydrogen bonds—or a pleated beta sheet—lateral connections of the polypeptide chain formed by hydrogen bonds between parallel regions.

Tertiary structure results from the folding of alpha helices and beta sheets into a compact globular structure, this shape is largely responsible for the function of the protein.

Quaternary structure is the overall structure of the protein, formed by the aggregation of two or more polypeptide chains.

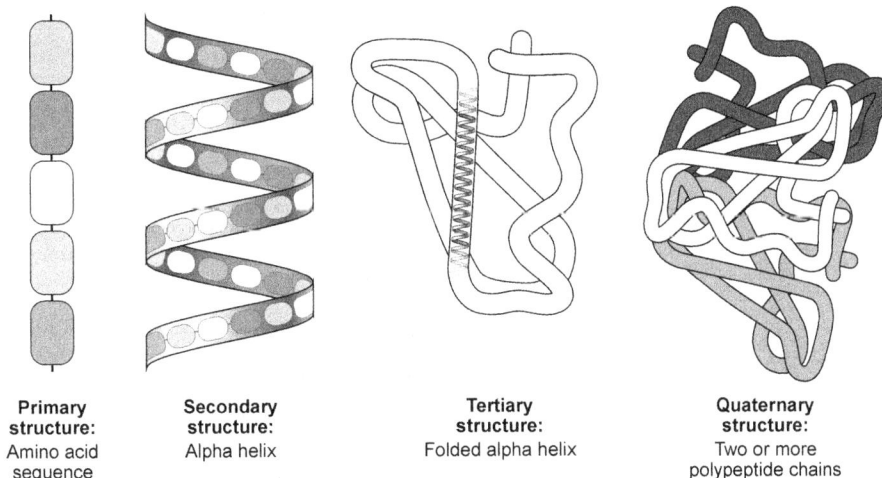

| Primary structure: Amino acid sequence | Secondary structure: Alpha helix | Tertiary structure: Folded alpha helix | Quaternary structure: Two or more polypeptide chains |

4.A.2: The structure and function of subcellular components, and their interactions, provide for essential cellular processes.

The cell is the basic unit of all living things. The two major types of cells are prokaryotes and eukaryotes. Prokaryotes are the most numerous and widespread organisms on earth. Bacteria were most likely the first cells and date back in the fossil record to 3.5 billion years ago. Their ability to adapt to the environment allows them to thrive in a wide variety of habitats. While there are several differences between cells, as will be discussed in detail below, all cells contain a cell membrane, DNA, ribosomes, and cytoplasm.

Anatomy and function of the cell

Prokaryotes do not have a defined nucleus or a nuclear membrane. Bacteria are examples of prokaryotes, and have freely floating DNA, RNA, and ribosomes in an area of the cell called the **nucleoid**.

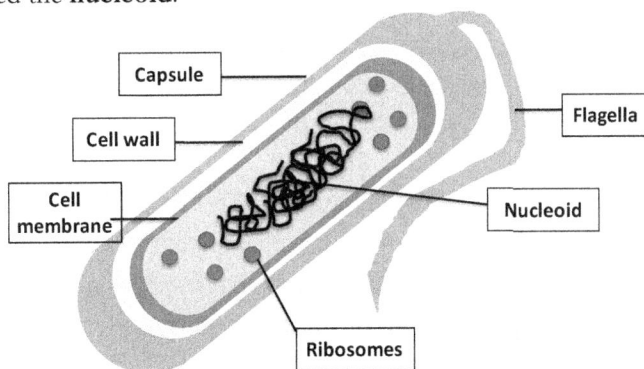

Anatomy of a prokaryotic cell

For the most part, **eukaryotes** are larger than prokaryotic cells and make up protists and multicellular fungi, plants, and animals. Eukaryotic cells maintain internal membranes that partition the cell into specialized regions called **organelles**, all with specific functions.. Cytoplasm inside the cell supports the organelles and contains the ions and molecules necessary for cell function.

Eukaryotic cells also maintain internal membranes that partition the cell into specialized regions so that cell processes can operate with optimal efficiency by increasing beneficial interactions, decreasing conflicting interactions, and increasing surface area for chemical reactions to occur. Each compartment or membrane-bound organelle localizes reactions, including energy transformation in mitochondria and production of proteins in rough endoplasmic reticulum.

The most significant differentiation between prokaryotes and eukaryotes is that eukaryotes have a **nucleus**. The nucleus is the "brain" of the cell that contains the genetic information in the form of DNA, which is organized into chromosomes. The nucleus is the site of transcription of the DNA into RNA. The nucleus contains a structure called the **nucleolus**, responsible for making rRNA and assembling ribosomes. A nuclear envelope surrounds the nucleus that contains many pores that let RNA out of the nucleus.

Ribosomes are responsible for protein synthesis for use in the cell or to be secreted. Ribosomes are made up of ribosomal RNA and proteins and have two subunits that come together for protein synthesis. They can be found free floating in the cytoplasm or attached to the endoplasmic reticulum.

The **endoplasmic reticulum** (ER) is the "roadway" of the cell and transports materials that will be secreted or used within the cell. There are two types of ER: smooth and rough. Rough endoplasmic reticulum (RER) contains ribosomes on its surface and aids in the synthesis of proteins that are membrane-bound or destined for secretion. Smooth endoplasmic reticulum (SER) does not contain ribosomes and is the site of hormone, lipid, and steroid synthesis. Many of the products made in the ER proceed to the Golgi apparatus.

The **Golgi apparatus** consists of a series of flattened membrane sacs in which molecules that are made in other parts of the cell (like the ER), are sorted, modified, and packaged. Vesicles are produced by the golgi bodies containing products so they can travel to the plasma membrane. Golgi bodies also contribute to the production of lysosomes.

Lysosomes are small, membrane-bound sacs that contain digestive enzymes functioning at an acidic pH. Lysosomes break down damaged or old cell components, unnecessary substances, viruses, and particles that have been ingested.

Mitochondria are large organelles that are the "powerhouses of the cell." Cellular respiration, the process of ATP production that supplies energy to the cell, takes place in the mitochondria. Muscle is very densely laden with mitochondria because muscle requires a large amount of energy. Mitochondria have their own DNA, RNA, and ribosomes; they also have two membranes: a smooth outer membrane and a folded inner membrane. The folds inside the mitochondria are called **cristae**—important for increasing the surface area and containing enzymes important for ATP production. The production of ATP results in the formation of a proton gradient across the inner mitochondrial membrane. Therefore, the double membrane provides a separation between a region of high proton concentration and a region of low proton concentration.

While mitochondria produce energy by cellular respiration, **chloroplasts** are the organelles in plant cells in which the electron transport chain reaction occurs during photosynthesis. The chloroplast has an inner membrane space called the stroma that encloses sacs called **thylakoids** that contain the photosynthetic pigment chlorophyll. The chlorophyll traps sunlight inside stacks of thylakoid called grana to generate ATP, which

Mitochondria specialize in energy capture and transformation.

is used in the stroma to produce carbohydrates and other products. **Plastids** are found only in photosynthetic organisms. They are similar to mitochondria in that they both have a double membrane structure. They also have their own DNA, RNA, and ribosomes and can reproduce if they need to increase their ability to capture sunlight. There are several types of plastids. The **chromoplasts** make and store yellow and orange pigments. They provide color to leaves, flowers, and fruits. The **amyloplasts** store starch and are used as a food reserve. They are abundant in roots like potatoes.

The Endosymbiotic Theory states that mitochondria and chloroplasts were once free living and possibly evolved from prokaryotic cells. At some point in our evolutionary history, they entered the eukaryotic cell and maintained a symbiotic relationship with the cell. The fact that both types of cells have mitochondria and chloroplasts that have their own DNA, RNA, ribosomes, and are capable of reproduction supports this theory.

The following is a diagram of a generalized animal cell:

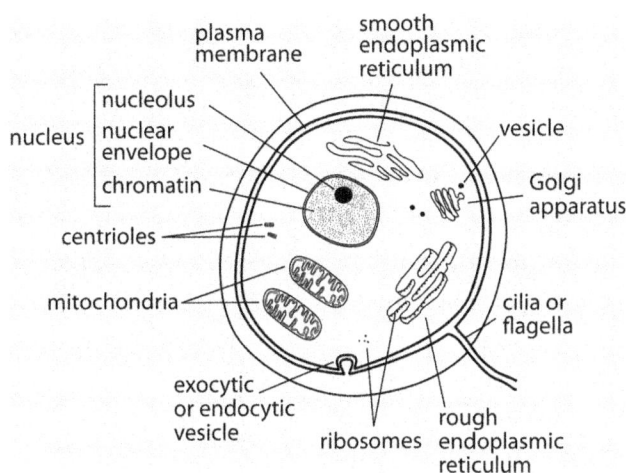

A generalized animal cell

Chloroplasts are specialized organelles found in algae and higher plants that capture energy through photosynthesis.

To summarize, below is a chart comparing different cells types:

	Prokaryote	Eukaryote	
		Plant	Animal
Cell Wall	√	√	X
Cell Membrane	√	√	√
Ribosomes	√	√	√
Nucleus	X	√	√
Organelles	X	√	√

Additional anatomy of the cell

The **cytoskeleton**, found in both animal and plant cells, is composed of protein filaments connected to the plasma membrane and organelles. The cytoskeleton provides a framework for the cell and aids in cell movement. Three types of fibers make up the cytoskeleton:

- **Microtubules** are the largest of the three fibers and are composed of the protein tubulin. Cilia and flagella are made up of microtubules and are used for locomotion;

for example, in sperm cells, and cilia that line the fallopian tubes and trachea. Centrioles, which form the spindle fibers that pull the cell apart into two new cells during cell division, are also composed of microtubules. Centrioles are not found in the cells of higher plants.

- **Intermediate filaments** – Intermediate in size, they are smaller than microtubules, but larger than microfilaments. They help the cell keep its shape.
- **Microfilaments** are the smallest of the three fibers and are composed of actin and small amounts of myosin (like in muscle tissue). They function in cell movement like cytoplasmic streaming, endocytosis, and amoeboid movement. They also aid in pinching the two cells apart after cell division, forming two new cells.

4.A.3: Interactions between external stimuli and regulated gene expression result in specialization of cells, tissues and organs.

Communication

Cells communicate by sending and receiving signals and have evolved a variety of signaling mechanisms to accomplish the transmission of important biological information. The transmission of signals by molecules from outside of the cell to the inside is initiated by cell-surface receptors during the process called signal transduction, as described in section 3.D.3.

In single-celled organisms, signaling influences the response of the cell to its surroundings. For example, pheromones can trigger developmental or reproductive actions by the cell; signals can also influence how the cell moves or how cells move in relation to each other to control the population density.

Chemicals can help send signals from cells that are close to each other. A signal can be a molecule or a physical or environmental factor. Other times the signal works by interacting with receptor proteins that contact both the outside and inside of the cell. In this case, only cells that have the correct receptors on their surfaces will respond to the signal. Membrane signaling involves proteins called receptors embedded in the cell's membrane that biophysically connect the triggers in the external environment to the ongoing dynamic chemistry inside a cell. Signaling at the membrane also involves ion channels, which allow the direct passage of molecules between external and internal compartments of the cell.

As an example, light is translated into chemical messengers inside the cone and rod cells of the retina by receptors that allow ion currents to flow in response to photons—this allows us to see! In addition, chromatin structure (and therefore gene expression) is affected by growth factors that interact with the cell membrane. This results in receptor regulation during development which guides the path of migrating cells to ultimately control how an entire organism is wired together.

Differentiation in development is due to external and internal cues that trigger gene regulation by proteins that bind to DNA.

4.A.4: Organisms exhibit complex properties due to interactions between their constituent parts.

Osmoregulation and excretory systems

Animal metabolism produces plenty of waste products which need to be excreted from the body. There are many specialized organs and cell types found in all animals that help in this process. For example, earthworms have a long coiled tube structure called the nephridial tubule, which has a comparative function similar to that of a vertebrate kidney. Kidneys are the major excretory organ of humans but are not limited to removal of metabolic waste. Kidneys and other excretory organs also eliminate excess water, ions, or other non-metabolized compounds that are taken in with food and drink. They function to maintain **osmoregulation**—balance the body fluids, waste extraction, and ions in the body and keep the body at **homeostasis**—balancing the concentrations of water and ions in the body fluid. If the ion concentration in the body is too high, desiccation occurs and cells shrink. If the water concentration is too high, then cells will swell and dilute the ions. This process of water movement between low ion concentration into high ion concentrations is called osmosis, as we previously discussed. In a physiological system, this osmoregulation occurs in two steps; first, the organ filters the blood (in the example of a kidney it would be blood) into a tubule. Second, as the filtrate passes through the tubule, the needed molecules are pumped out and return to the body fluid, while metabolic waste, excess water, and other molecules in the body fluids are pumped into the filtrate. The resulting fluid, urine, is eliminated through the open end of the tubule outside the body, the urinary tract.

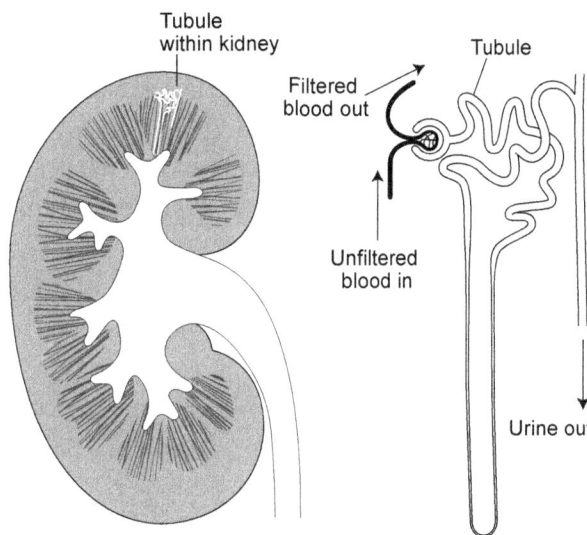

Tubule within kidney

Filtered blood out

Unfiltered blood in

Tubule

Urine out

Nerve signal transmission and response to external stimuli

We previously discussed how the neurons communicate using ionic gradients to generate action potentials, release neurotransmitters across gap junctions, and maintain physiological readiness. Neurons responding to extrinsic stimuli are using the same mechanisms. Stimuli first generate local gradient potentials in the sensory organ to elicit two types of responses. First, is an excitatory response: neurotransmitters are released and then bound by neuronal dendrites causing the cell to become permeable, and resulting in an influx of sodium ions through channels located near the binding site. The second type of response is an inhibitory response which prevents postsynaptic potential. Usually this is caused by GABA released into the synapse to prevent other neurotransmitters from causing an excitatory response.

4.A.5: Communities are composed of populations of organisms that interact in complex ways.

A **population** is a group of individuals of the same species that live in the same area and interact with each other. These interactions are commonly reproducing and competing for food and space (shelter, territory). **Demography** is the study of the size and structure of populations across space and time. Population size is affected by four processes that occur simultaneously: birth, death, immigration, and emigration. Immigration is the influx of new individuals coming from adjacent habitats, whereas emigration is when living individuals leave the population. **Population growth** is positive when the sum of births and immigration are higher than the sum of deaths and emigration.

Perhaps with the exception of chemoautotrophic bacteria, which obtain energy by the oxidation of organic and inorganic compounds, no single individual is capable of existing alone. All species play a role as prey and predators, and as hosts and parasites. All species benefit, directly or indirectly of the presence of another species or by the habitats created by them. Interactions in nature can be of many types, and are generally called as **symbiotic relationships** or **symbiosis** (from the Greek "living together"). Usually symbiosis is classified in three ways: When an interaction between two species is beneficial to both species, it is called **mutualism**. When only one species benefits and the other one is unaffected, it is called **commensalism**. **Parasitism** is an interaction where there is one beneficiary, the parasite, and the host species is harmed. Parasitism can lead to the host's inability to absorb nutrients and to reproduce. Parasite-host interactions are usually non-lethal, as the parasite can only benefit from its host when it is alive, so due to co-evolution, non-lethal relationships tend to be more stable. **Parasitoidism** is a special exception, where the parasite kills its host in one stage of its life cycle. Mutualism, commensalism, and parasitism are widespread in nature. In **amensalism,** one species is harmed while the other species is not negatively affected, but it does not necessarily profit from the interaction. **Antibiosis** is an example of amensalism, where one species produces chemical substances that, when released in the environment, may kill or inhibit the growth of others. Antibiotics, such as penicillin and rifamycin, are examples of such substances.

Predation is an asymmetric interaction where one species profits, the predator, while the other is harmed or killed. Predators that feed on primary producers are called **herbivores**, also called primary consumers. Species that feed on animals are called **carnivores**, and carnivores usually occupy the highest trophic levels. (The trophic level is the position that an organism occupies on the food chain) Species that feed from more than one trophic level, e.g. eating animals and plants, are called **omnivores**. There are also organisms called **detritivores**, that obtain their nutrients by eating portions of decomposing plants and animals. Predators can regulate the population size of their prey through direct and indirect effects. Direct effects are those consequences of death or injury caused by the predator. Indirect effects occur mainly through predator-induced changes in behavior, which ultimately affect the individual's capacity to obtain energy, shelter, or to reproduce. For instance, a small mammal may choose to stay hidden in a shelter instead of going out to forage for food, because it perceives the presence of

Mathematical or computer models are used to illustrate and investigate population interactions within and environmental impacts on a community.

predators nearby. Furthermore, the population size of the predator is also affected by prey availability.

Types of species interactions in nature

Species in the interaction	Mutualism	Commensalism	Parasitism	Amensalism	Predation
Species A	+	+	– (host)	–	– (prey)
Species B	+	=	+ (parasite)	=	+ (predator)

(+) benefit; (-) harm; (=) no effect.

Some species-species relationships can transition between the three types of symbiosis over the lifetimes of the organisms. For example, acacia trees can have a colony of ants living in its thorns. The ants provide protection from herbivorous insects and the plant houses and feeds the ants. However, if the population of ants becomes too big, the tree can suffer from the needs of the colony.

4.A.6: Interactions among living systems and with their environment result in the movement of matter and energy.

Earth is composed of five major compartments: biosphere, geosphere, lithosphere, hydrosphere and atmosphere. Through the biogeochemical cycles, elements that are essential to life flow from one compartment to another by alternating into organic and inorganic forms. Due to the important roles in global warming and agriculture, extensive study has been performed on the carbon and nitrogen cycles, respectively.

The carbon cycle

The largest carbon pool on Earth is found in the lithosphere, and consists mainly of carbonates. The second largest is found in the hydrosphere, in inorganic forms in the oceans. In nature, carbon in the form of CO_2 is produced by autotrophic and heterotrophic respiration, and enters the atmosphere. It is taken from the atmosphere during photosynthesis, and transformed in various organic molecules, which will form the components of the biosphere. When organisms die, their tissues are eventually consumed by detritivores and decomposers, and through their metabolism, organic carbon will be lost to the atmosphere in the form of CO_2.

The carbon cycle plays a major role in the greenhouse effect. The greenhouse effect is the warming of Earth's surface by the radiation energy emitted by the greenhouse gases in its atmosphere. CO_2, methane, water vapor, nitrous oxide, and ozone are all greenhouse gases contributing to global warming.

Nitrogen cycle

While most of Earth's carbon is stored in the lithosphere, the largest pool of nitrogen is found in the atmosphere, as nitrogen gas (N_2). Although nitrogen is an essential macro element to all life forms, its cycling and availability depends largely on specialized bacteria. N_2 is a very stable molecule (the two N atoms are bound together by triple bonds), and its transformation into organic forms is only accomplished by specialized bacteria called nitrogen fixers, or **diazotrophs**. Bacteria of genera *Rhizobium* and *Azotobacter* are examples

Energy flows, but matter is recycled.

Changes in regional and global climates and in atmospheric composition influence patterns of primary productivity.

of nitrogen fixers. They possess an enzyme called nitrogenase, which combines water with N_2 into ammonia (NH_3), which is later transformed into organic compounds by bacteria. The transformation of ammonia into nitrite and nitrate is called nitrification. Plants can absorb nitrogen from water and soils when present in the form of nitrite (NO_2), nitrate (NO_3), or ammonium (NH_4^+). When organisms die, the nitrogen in their organic molecules will be transformed back to an inorganic form, such as ammonium (NH_4), and eventually released in the water or soil, where it can be absorbed by plant roots. This is called ammonification. Under anaerobic conditions, such as in dump soils, bacteria can transform nitrate back to nitrogen gas through denitrification, which reaches the atmosphere and completes the nitrogen cycle.

Trophic ecology, food webs and food

Within the biosphere, elements and energy move along trophic levels, which, you remember, are their positions on the food chain. All living organisms are either producers or consumers. Autotrophs, or producers, constitute the first trophic level and form the basis of all food chains, as they turn inorganic material into organic carbon, using the energy from the sun. The total amount of carbon taken from the atmosphere by producers is called **Gross Primary Production** (GPP). Subtracting the amount of CO_2 produced during respiration from the GPP yields the **Net Primary Production** (NPP), so NPP = GPP – respiration. NPP varies widely among biomes, as it depends on a combination of temperature, precipitation, nutrient availability, daylight, etc. For instance, while in a tropical forest NPP is about 17.8 Petagram (Pg) C per year, in deserts and tundra NPP is less than 1 Pg C per year. For reference, a petagram is equal to one billion metric tons!

All other trophic levels will use the energy fixed by producers in their own metabolism. Herbivores and detritivores are called primary consumers, and occupy the second trophic level. Animals who feed on these two groups are carnivores, and called secondary consumers (third trophic level). Tertiary consumers are carnivores that feed on primary and secondary consumers, and are called top-predators, or apex predators (fourth trophic level). Secondary and tertiary consumers can also be omnivores. These relationships are summarized in a table below.

Trophic level	Group	Examples	Feed on
First	Producers	Plants, cyanobacteria, metanogens	---
Second	Primary consumers	Herbivores, detritivores, decomposers	Producers
Third	Secondary consumers	Carnivores and omnivores	Primary consumers and producers
Fourth	Tertiary consumers	Carnivores and omnivores	Primary and secondary consumers

As seen above, all species in a community are connected, directly or indirectly, through trophic interactions. For that reason, any alteration in the environment eventually affects all species that inhabit it. One of the main threats to many ecosystems worldwide is pollution derived from human activities. Two processes, bioaccumulation and biomagnification, or even a small or localized event, may affect species living far away.

Human activities impact ecosystems on local, regional and global scales.

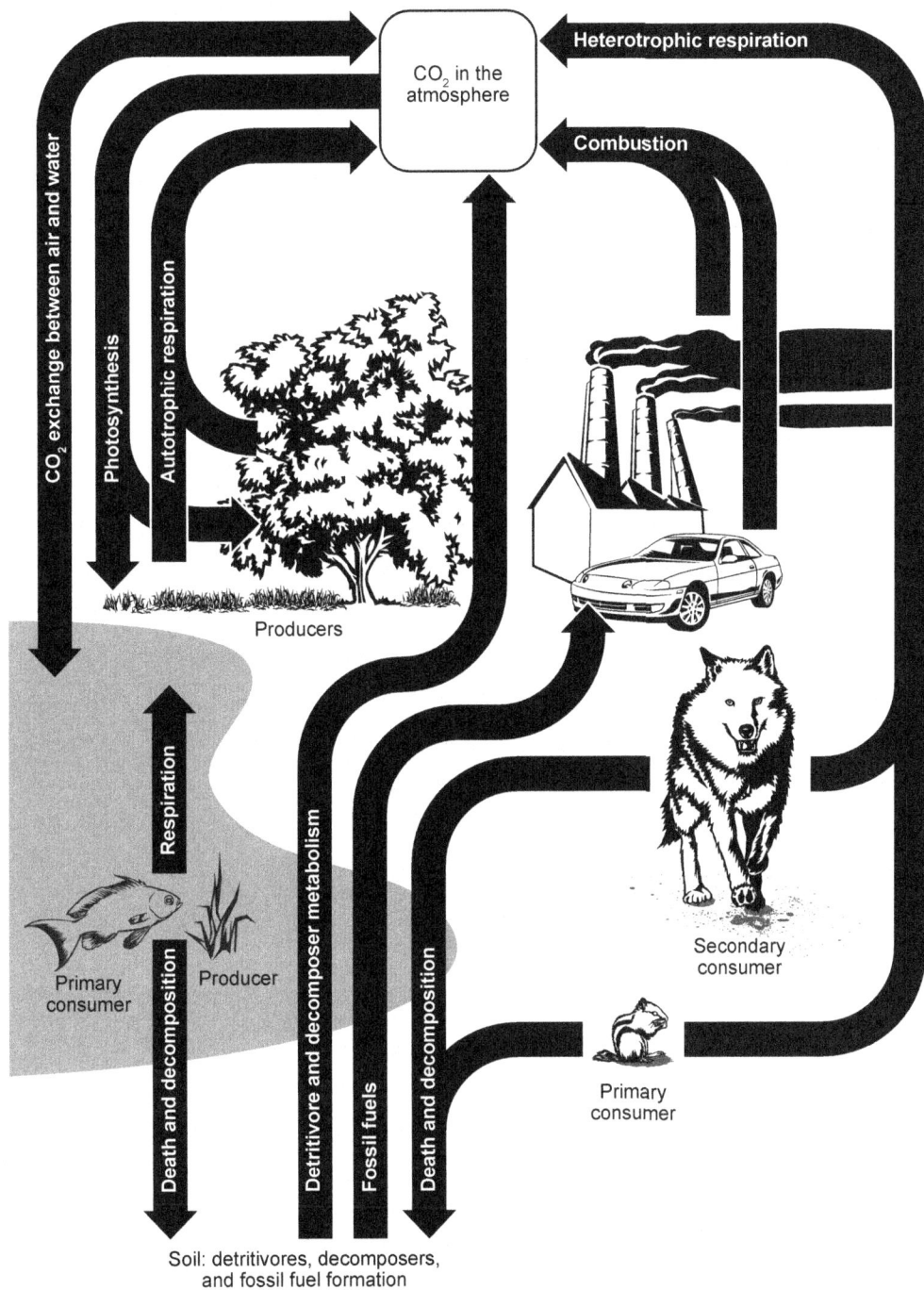

Summary of ecosystem interactions

Consider, for instance, a lake with waters contaminated by mercury. In this lake, aquatic plants would ingest mercury along with the water absorbed through the roots. Through a series of metabolic processes, this mercury would eventually reach, and be stored in the leaves. These leaves are eventually consumed by herbivores, and increased consumption will lead to increased mercury concentrations in tissues. The increase in contamination levels during the life of an individual is called **bioaccumulation**. When secondary consumers (a carnivore fish) feed on primary consumers (aquatic invertebrate herbivores), they will, in turn, accumulate the contaminant. As each individual, in a given trophic level, feeds on many individuals of the lower level, the contamination levels will increase from one trophic level to the next. This process is known as **biomagnification**.

For example, fish can eventually disperse to adjacent lakes, where they are prey for larger fish, mammals, and birds. Some of these fish, mammals, and birds may be hunted, and sold for human consumption in cities nearby. This is how the harmful effects of pollutants can spread within and across different ecosystems.

A well-known case of biomagnification is the insecticide **DDT** (**dichlorodiphenyltrichloroethane**). From the 1950s through the 1980s, DDT was sprayed over agricultural fields and in cities, to control the populations of disease vectors. Although DDT was produced and used only to eliminate insect pests research later revealed its harmful effects on birds and mammals, including humans, through biomagnification. Nowadays, DDT production is prohibited in most countries, but their effects will persist in the environment for decades.

Enduring understanding 4.B: Competition and cooperation are important aspects of biological systems _____

4.B.1: Interactions between molecules affect their structure and function.

The formation and breakdown of chemical bonds to create various compounds required for life is facilitated by **enzymes**, which are biological catalysts capable of speeding up chemical reactions.

The rate of reactions is based on the **activation energy**, that is, the minimum requirement of energy necessary to allow a chemical reaction to occur. Since enzymes accelerate reactions, they are capable of lowering the activation energy, as shown in the figure below:

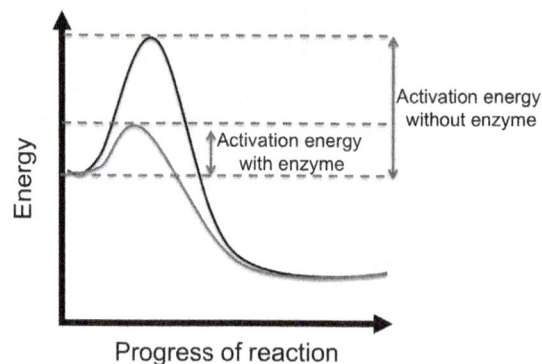

In general, there are specific enzymes for a particular reaction, and the reactants are called the **substrates**. These target molecules, the substrates, are bound by the enzymes to the active site forming an **enzyme-substrate complex**. Interestingly, when binding to the substrates, enzymes have a mild change to their shape (termed **induced fit**), in which they conform to the shape of the substrates. The result of this complex is the formation of a product, after which the enzyme is free to facilitate another reaction.

The change in function of an enzyme can be interpreted from data regarding the concentrations of product or substrate as a function of time. These representations demonstrate the relationship between an enzyme's activity, the disappearance of substrate, and/or presence of a competitive inhibitor.

Sometimes enzymes require a cofactor or coenzyme to perform its job. Interaction with a cofactor or coenzyme can be crucial to catalyzing the reaction since they allow the enzyme to become active.

Similarly, enzyme activity can be regulated. **Allosteric sites** are locations on the enzyme which can bind an allosteric inhibitor or activator. As you can imagine, binding of an allosteric inhibitor will prevent the enzyme from taking its active form, and the allosteric activator promotes the activation of the enzyme, and thus catalysis of the reaction. The concentration of available substrate or product can also regulate enzyme activity.

In addition, chemical substances can be used to regulate enzymes in one of two ways:

Competitive inhibition, in which the substance is able to bind to the active site of the enzyme, therefore rendering the actual substrate unable to bind for reaction progression.

Noncompetitive inhibition is when a substance binds to another site on the enzyme, not the active site. This results in an alteration to the enzyme such that the substrate cannot bind and therefore cannot activate the reaction.

4.B.2: Cooperative interactions within organisms promote efficiency in the use of energy and matter.

At the cellular level, the plasma membrane, cytoplasm and organelles contribute to the overall specialization and function of the cell. In multicellular organisms the different functions of specialized cells fulfill distinct needs for the organism as a whole through cooperative interactions. In the lungs for example, cells that line the alveoli increase the permeability of gasses through tissues in order to be dissolved in the blood. At the same time, alveoli also transport carbon dioxide out of the blood while keeping the blood inside the circulatory system. The circulatory system transports these gasses throughout the body and oxygenates the tissues at the cellular level, while removing waste and expelling carbon dioxide that is a result of cellular metabolism. The waste products are removed by complementary cells of the kidney and other excretory organs when the blood passes through the tissue. This maintains homeostasis for metabolic activity, ions, water, and metabolic waste. This is achieved through the cooperative interactions of highly specialized cells.

Cellular cooperation is not limited to individual cells of the organism. In deep sea hydrothermal vents, for example, bacterial communities colonize the gut and gills of the animals living there. These bacteria harvest resources from the only source of energy, the sulfur vents. They are able to convert this into energy for the animal in return for the animal providing a niche in which the bacteria can grow. This symbiotic relationship is at the cellular level, but in the end, it is sustaining the ecosystem.

Organisms have areas or compartments that perform a subset of functions related to energy and matter, and these parts contribute to the whole.

4.B.3: Interactions between and within populations influence patterns of species distribution and abundance.

Community ecology addresses the interactions between coexisting species, and between the species and their resources. It also investigates how biotic and abiotic factors determine the species composition of a given community.

There is a finite number of resources available in communities. When two species compete for the same resources they are competing in the same niche. The niche cannot support both species if one of the species has a competitive advantage of acquiring the resource. The species that has adapted with the greater advantage will outcompete the other species and eventually can replace it or cause localized extinction within the community. This is called the **competitive exclusion principle**, (also known as *Gause's Principle*); no two species can sustain coexistence if they occupy the same niche.

Coexistence can occur when two species occupy different or even slightly different niches which can sometimes be difficult for scientists to distinguish.

Twelve species of guppies coexist in the same river by feeding on organisms in different depths and by using different feeding behaviors to obtain food. Natural selection can act to diverge these traits, diversifying the features that reduce niche competition, and can even lead to **character displacement** or **niche shifting.** This is best exemplified in Darwin's finches that had varied beak sizes, allowing particular species to specialize in eating seeds of particular sizes.

As the environment is highly heterogeneous due to gradients of abiotic factors, populations are seldom evenly distributed across space and time because individuals are always seeking more favorable habitats, with fewer predators and/or more resources. Two things often occur:

> **Dispersion** — individuals find new habitats randomly

> **Migration** — a phenomenon where a population, or a part of it, moves towards another specific habitat to search for food or mates.

Population density is the number, or biomass, of individuals per unit area or volume. The distribution of individuals in an area is called dispersion. Dispersion patterns can be clumped, when individuals are grouped in more favorable patches, and less favorable ones are empty or uniform, when density across patches is similar or random.

Each organism can only exist within a set of biotic and abiotic characteristics. This set is known as the species' **niche**. The modern concept of the niche is known as the Hutchinsonian Niche, which defines it as an n-dimensional hypervolume, where each dimension represents the range of one of the resources used by the species, and n is the number of resources. Similar species within a community may display some degree of niche overlap, which is when two species compete for the same resource, and therefore develop similar traits. Two species with overlapping niches can coexist in a community through **resource partitioning** which is when two similar species may be active at different times of the day, or forage in different microhabitats of a shared patch in order to exist together.

Any organism, if free of intraspecific or interspecific competition, could use all the resources of its **fundamental niche** which contains the environmental conditions that allow the species to exist. In nature, however, competition is a strong force, and restricts each organism to the smaller fraction of its fundamental niche called the **realized niche.** For instance, consider that in a forest patch there are ten plant species that could provide nectar to a butterfly species A which is capable of using all ten species for food, therefore part of the fundamental niche. However, there are two other butterfly species (B and C) in the patch, and they compete strongly with species A. Species B and C have a strong preference for five of the ten plants, which makes it difficult for species A to use them. As a result, species A only uses the nectar from five out of the ten plants present in the patch. These five plants are part of the realized niche of species A.

A population of organisms has properties that are different from those of the individuals that make up the population. The cooperation and competition between individuals contributes to these different properties.

The niche of a species also defines the functions it will perform in the ecosystem. Examples of functions are primary production, decomposition, predators that control prey populations, and so on. **Keystone species** affect their ecosystems and the other species in it, on a larger scale than it would be expected based on their biomass. Some keystone species are also **ecosystem engineers**: they modify, create and destroy habitats, largely affecting all species related to them. Beavers, ants, earthworms, elephants, and prairie dogs, are examples of ecosystem engineers. **Pioneer species** are examples of ecosystem engineers as well: they can colonize empty or disturbed places before any other living organism. After they grow, their biomass serves as food for subsequent species, and/or as a substrate for colonization. Lichens and mosses that can grow on bare rocks are examples of pioneer species.

When a keystone or ecosystem engineer species is lost or removed, drastic and rapid changes are seen in the community. The interaction between sea urchins, otters and kelp forest illustrate the important role of the keystone species. Kelp forests are formed by brown marine macroalgae in shallow areas. They are very productive, meaning that they play an important role in the marine uptake of CO_2 from the atmosphere, and they serve as food and habitat for a diverse group of animals. Sea urchins graze on kelp, decreasing their biomass where they are abundant. Sea otters, on the other hand, prey on sea urchins and control their populations. Where otter populations have been depleted, sea urchins flourish and the kelp forest disappear, creating barren landscapes unsuitable for life. By the role they play in controlling the sea urchin populations and securing the kelp forests, the otters are considered as the keystone species.

4.B.4: Distribution of local and global ecosystems changes over time.

Since the formation of the Earth, the landscapes have changed due to a combination of geological, climatic, and biological processes. **Mass extinctions** are events where a large number of species becomes extinct in a short period of time (on a geological scale, of course). After a mass extinction, the species composition and the landscapes of the affected ecosystems change completely: As predicted from the **Niche Theory**, when species are extinct, an empty niche is left, giving an opportunity for the emergence and occupation of other species. For example, the event that led to the extinction of dinosaurs (Cretaceous-Paleogene) may have provided the opportunity for the great radiation of modern-day mammals and birds. Through the fossil records, scientists identified five mass extinction events on Earth (see table below). There are three main causes for past mass extinctions: flood basalt, sea-level falls, and asteroid impacts. The first two are more common, and probably caused most of the Earth's extinction events. Asteroid impacts are only directly associated with the Cretaceous-Paleogene extinction. The impact of this meteorite left a crater in the Yucatan peninsula of Mexico.

Geological and meteorological events impact ecosystem distribution, and human impact accelerates change at local and global levels.

Time				
450–440 Ma	375–360 Ma	252 Ma	201 Ma	66 Ma
Ordovician-Silurian	Late Devonian	Permian–Triassic	Triassic–Jurassic	Cretaceous–Paleogene

The continental drift, caused by the movement of tectonic plates, also played an important role in shaping the landscape of the Earth. As the continents separated and drifted to new regions of the globe, they were exposed to new abiotic conditions, including variations in temperature and sunlight. Think of the lower temperatures and shorter days at the poles, and warmer and drier conditions at the Equator. These factors shaped the characteristics of the species living in these regions.

Climate change also contributes to change in species composition and abundance of some ecosystems. For example, "El Niño" is a phenomenon that repeats every 3 to 4 years on average. It causes disturbances in South America and in the Pacific, usually increasing precipitation in the former, and droughts in the latter. The "El Niño" effect is known to affect fisheries in Chile and Peru.

Biotic processes, such as contact with new species, and the pathogens they carry also alter ecosystems and populations. For example, the "Black Death" killed 70 to 200 million Europeans between 1343 and 1353. Other plagues affected plant species as well. The "Dutch elm disease" caused the death of almost all Elm tree populations in certain countries of Europe, and also affected the United States. This disease is caused by a fungus (Ascomycota) carried by the elm bark beetle. It originated in Asia and spread towards the northern temperate forests in the early 1900s. Similarly, the Emerald ash borer, a beetle from Asia that feeds on ash trees, is harmless in its native region due to the low population size. However, after the Emerald ash borer introduced in the United States and Canada they became a threat to native ash populations.

Enduring understanding 4.C: Naturally occurring diversity among and between components within biological systems affects interactions with the environment _____

4.C.1: Variation in molecular units provides cells with a wider range of functions.

Variations in lipid membranes

Variations in lipid composition of cell membranes of many species appear to be adaptations to specific environmental conditions. For example, the ability of the cell to change its lipid compositions in response to temperature changes has evolved in organisms that live where temperatures vary. Functions of membrane proteins vary due to the many types of proteins that decorate it. They often group together and are embedded in the fluid matrix of the lipid bilayer. The different types of proteins determine most of the membrane's specific functions, such as ion transportation, secretion, signaling and recognition. The three types of proteins found in lipid membranes are, **peripheral proteins** which are bound to the surface of the membrane, **integral proteins** which penetrate the hydrophobic core, and integral proteins that span the membrane are called **transmembrane proteins**.

Variations within molecular classes provide cells and organisms with a wider range of functions.

Peripheral protein

Integral protein

Transmembrane protein

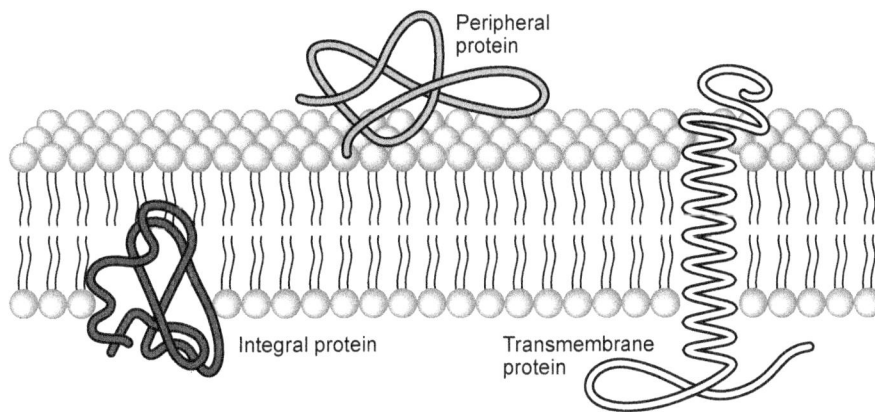

Variation in amino acid composition results in diverse protein function. For example, the hydrophobic regions of an integral protein consist of one or more stretches of nonpolar amino acids, often coiled into alpha helices. This is the case for hemoglobin (Hb) variants which give us our blood types, (example, Hb A = type A). It is also the case for plants' chlorophyll to produce chlorophyll *a* and chlorophyll *b*.

Variation in antibodies and the immune system

While we previously discussed immunity, there are variations in the types of immune responses after exposure to an antigen: humoral and cell-mediated.

1. **Humoral response** — Free antigens and antigen presenting cells activate B cells (lymphocytes from bone marrow) which transform into plasma cells that secrete antibodies. Memory cells are also generated that recognize future exposure to the same antigen. Antibodies defend the body against extracellular pathogens by binding to the antigen, providing an easy target for phagocytes to engulf and destroy. Antibodies belong to a class of proteins called immunoglobulins. There are five major classes of immunoglobulins (Ig) involved in the humoral response: IgM, IgG, IgA, IgD, and IgE.

2. **Cell-mediated response** – Infected cells activate T cells (lymphocytes from the thymus) which then bind to the infected cells and destroy them along with the antigen. T cell receptors on T helper cells recognize antigens bound to the body's own cells. T helper cells release IL-2, a cytokine that stimulates other lymphocytes (cytotoxic T cells and B cells) to participate in the immune response. Cytotoxic T cells kill infected host cells by recognizing specific antigens.

Both the humoral and cell-mediated responses are elicited by vaccines. Administration of very small amounts of antigen allows memory cells to recognize future exposure to the antigen, which results in the faster production of antibodies, and protection from the antigen!

Gene families and pseudogenes

A gene family is a group of genes that share important characteristics. They can have similar DNA sequences, provide instructions for making proteins with a similar structure or function, or dissimilar genes that are grouped together because the genes produce products that function together for a similar process. Gene family classification is principally used to understand how genes relate to each other, as some genes can be members of several gene families. For example, many innate immune genes are related to developmental signaling in early embryogenesis. Similarly, scientists can use information that is known about a gene in one family, and compare the sequences or products to predict the function of an unknown gene.

There are regions of DNA called **pseudogenes** that are characterized by a combination of homology to a known gene and nonfunctionality. The homology is implied by the similarities between the DNA sequences of the pseudogene and parent gene. They usually share a common ancestry and are a result of mutation, or duplication and mutation. However, the nonfunctionality is due to mutations that cause a breakdown in transcription, processing, translation, and/or protein folding. This can be caused by frameshift mutations or a new stop codon.

Think of it like this:

Gene: the fat dog ate his hat.

Frameshift mutation: the atd oga teh idh at.

Stop codon: the fat.

The first sentence makes sense, but deleting one letter and shifting everything over makes no sense, or adding a period in the middle loses the meaning. Pseudogenes are sometimes difficult to identify and characterize because although they look like known genes, they do not have a known function.

4.C.2: Environmental factors influence the expression of the genotype in an organism.

The environment plays an important factor in how genotypes are expressed as phenotypes. Though the genetic potential for a given trait may exist within a population, how the traits manifest is due in part to extrinsic forces that regulate the genotype. In humans, there is a genetic potential to reach a given height and weight, however this also depends on factors such as diet and exercise during development and throughout adulthood.

The genotype of a flower is the same, but extrinsic factors cause a responsive phenotype. In some reptiles the temperature can cause major developmental differences. For example, alligators lack sex-determining chromosomes, so their egg location in the nest determines the temperature to which they are exposed during development and subsequently, the sex. Eggs that are incubated at 34°C produce male offspring. The temperature causes androgenic hormones to be expressed in the developing embryo, which triggers the development of male offspring.

Similarly, animals that live in temperate zones demonstrate phenotypic responses in their fur to changing seasons and temperatures. Producing thicker coats in the winter that are more drab and light in color to blend into their environment while keeping them warm. The coats are thinned in the warmer months with colors that match the environment and help cool down the animal.

Similar effects are observed in traits such as pigmentation of certain species of hydrangea flowers depending on the acid or basic qualities of the soil, as measured by the pH scale. A soil pH of 5.5 or lower causes the flowers to have blue petals, while higher pH causes the petals to be pink.

An organism's adaptation to the local environment reflects a flexible response of its genome.

The **pH scale ranges** from zero to fourteen and is a reference to explain if a solution is an acid or base. A pH between zero and 6.9 is acidic. A pH between 7.1 and 14 is basic. We have provided a scale with examples of everyday items and their corresponding pH levels.

A high concentration of hydrogen ions (H^+) in solution constitutes an **acid.** For example, hydrochloric acid (HCl), stomach acid, has a pH of 2.0. **Neutral** solutions – neither acidic nor basic – are at 7 on the pH scale. A solution that releases hydroxide ions (OH^+), and not hydrogen ions, is a **base.** An example is sodium hydroxide (NaOH), which has a pH of 14.

The internal pH of most living organisms is close to 7. Human blood has a pH of 7.4 so it is quite neutral. Variation from this neutral pH can be harmful to the living organism. Biological fluids resist pH variation due to buffers that can accept or donate H^+ ions to or from a solution, and minimize the effects of H^+ and OH^- concentrations.

pH	Example
0	Battery acid
1	Stomach acid
2	Lemon juice
3	Vinegar
4	Acid rain
5	Black coffee
6	Milk
7	Distilled water
8	Seawater
9	Baking soda
10	Milk of magnesia
11	Ammonia
12	Soapy water
13	Oven cleaner
14	Drain cleaner

The pH of a substance has a dramatic effect on the environment as well. Acidic precipitation (rain, snow, or fog) is caused by sulfur oxides and nitrogen oxides in the environment that react with water in the air to form acids that fall down to earth as precipitation, commonly called "acid rain." In addition, a change of pH in the environment can affect the solubility of minerals in the soil, resulting in stunted forest growth.

4.C.3: The level of variation in a population affects population dynamics.

Genetic variation in a population is a very important factor in the long-term survival of a population. Increased genetic diversity improves the possibility of adapting to extrinsic changes including, new niche opportunities, altered predatory patterns, and developing resistance to disease. This is achieved by the maintenance of large populations, migration and gene flow between populations. However, certain events can drastically reduce the size of a population, and this is called a **population bottleneck**. The bottleneck can be caused by numerous factors including habitat destruction, disease, over hunting, or

Population ability to respond to changes in the environment is affected by genetic diversity. Species and populations with little genetic diversity are at risk for extinction.

loss of a niche. The remaining population has lower genetic diversity and fewer genetic characteristics to allow for adaptation and diversification within the remaining niches. One example of this is the cheetah. Though the large cat is sustaining larger populations in the wild, genetic analysis of cheetah populations show that there is very little genetic diversity. This indicates that there was a large loss of genetic variation. Though the cheetah is successful now, the remaining population faces a higher level of genetic drift. In small populations, infrequently occurring alleles face a greater chance of being lost, which can further decrease the gene pool. This refers back to our discussion of allele frequency in Big Idea 2.

4.C.4: The diversity of species within an ecosystem may influence the stability of the ecosystem.

Ecosystems have two important properties, resistance and resilience. **Resistance** indicates how much disturbance an ecosystem can absorb without having its functions affected and depleted. **Resilience** is the time needed for an ecosystem to regenerate and return to a steady-state, after being affected by a disturbance. Disturbances can be biotic (insect outbreaks) or abiotic (floods, hurricanes, droughts). Both properties are important drivers of ecosystem stability. Ecologists have noticed that species-rich ecosystems show a higher stability in response to disturbances. However, the exact mechanism explaining this relationship is not fully understood. One hypothesis is that biodiversity could act as an "insurance" mechanism, i.e. in species-rich ecosystems there is a higher chance that at least one or a few species will have a combination of traits enabling them to withstand a disturbance.

Keywords

Molecule
Bonds
   Hydrogen
   Covalent
   Ionic
Electronegativity
Polarity
Condensation
Hydrolysis
Carbohydrates
   Monosaccharide
   Disaccharide
   Polysaccharide
      Starch
      Glycogen
      Cellulose
      Chitin
Lipid
Fats
Phospholipids
Hydrophobic
Hydrophillic
Amphipathic
Amino acids
Steroids
Proteins
   Primary/secondary/tertiary/
      quaternary structure
Hydrolysis
Prokaryotes/Eukaryotes
Organelles
   Ribosomes
   Endoplasmic reticulum
   Golgi appartus
   Lysosomes
   Mitochondria
      Cristae
Chloroplasts
   Thylakoid

Plastids
   Chromoplasts
   Amyloplasts
Cytoskeleton
   Microtubules
   Intermediate filaments
   Microfilaments
Osmoregulation
Homeostasis
Population
Demography
Population growth
Symbiosis
Parasitoidism
Mutualism
Commensalism
Amensalism
Antibiosis
Predation
   Herbivore
   Carnivore
   Omnivore
   Detritivore
Diazotrophs
Trophic level
Gross Primary Production
Net Primary Production
Bioaccumulation
Biomagnification
DDT
Enzymes
   Activation Energy
   Substrates
   Induced fit
   Allosteric site
Competitive/noncompetitive
   inhibition
Community ecology
Competitive exclusion principle

Character displacement/niche shifting
Population Density
Dispersion
Migration
   Niche
      Fundamental niche
      Realized niche
Resource partitioning
Keystone species
Ecosystem engineers
Pioneer species
Mass extinction

Niche Theory
Integral/Peripheral/Transmembrane
   proteins
Humoral response
Cell-Mediated response
Gene families
Pseudogenes
pH scale
Population bottleneck
Resistance
Resilience

Summary

Interactions between parts are paramount to the optimal function of all biological systems. Increased diversity and complexity in a biological system leads to increased capacity to respond to changes in the environment – and therefore greater chance of survival.

Molecules have subcomponents that determine the properties of the particular substances.

Cells have subcomponents called organelles, that interact to promote cell survival, growth, and reproduction.

Organs and organ systems interact to allow the organism to function as a whole.

Populations and ecosystems are equally affected by external and internal environmental factors that affect the physical and biological characteristics, keeping matter and energy in flux.

Competition and cooperation are important interactions in the maintenance of biological systems.

Variations in biological systems enable flexible responses to change: cells have more functions if there is variation in their molecules; and genetic diversity in populations allow for better survival of the species and the ability to respond to changing environmental conditions.

Chapter 4 Quiz

1. Monarch butterflies have an orange and black coloration that identifies them well known to predators. Since these butterflies eat milkweed, which is poisonous to other creatures, the predators know to stay away from them. The viceroy butterfly also has an orange and black coloration that is very similar to the monarch's, but the viceroy butterfly does not feed on milkweed, so it is not toxic. Even so, predators still avoid the viceroy. What type of adaptation keeps predators from eating the viceroy butterflies?

   A. Mutation

   B. Learning

   C. Reproductive isolation

   D. Mimicry

2. **Which of the following BEST explains why there are usually fewer than five trophic levels to most food chains?**

   A. Many primary consumers feed at more than one trophic level.

   B. The carrying capacity of the environment would be exceeded with more than five levels.

   C. Each trophic level only obtains a small fraction of the energy from the trophic level below it.

   D. The increased demand on the tertiary consumers would cause them to face extinction.

Question 3–5 refers to the food web in the diagram.

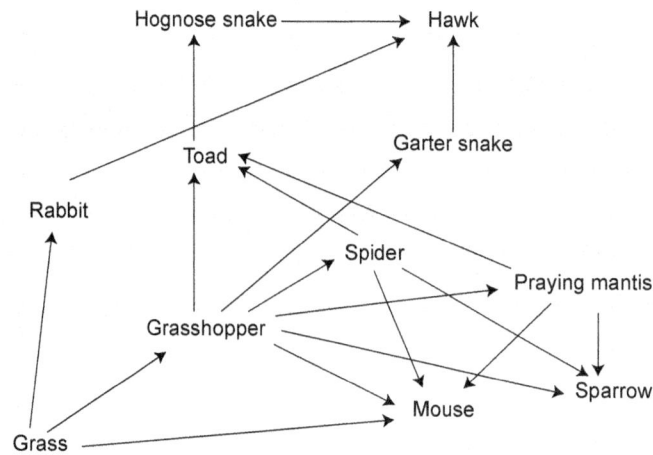

3. **In this food web, what would be a likely impact if the grasshopper population were to decrease?**

   A. The number of garter snakes would drastically decrease.

   B. The hawk population would decrease.

   C. The hognose snake population would increase.

   D. The mouse population would show a steady decline.

4. **In which sequence of events is the mouse considered a tertiary consumer?**

   A. Grass → grasshopper → spider → mouse

   B. Grass → rabbit → hawk → mouse

   C. Grass → mouse → hognose snake → hawk

   D. Grass → sparrow → praying mantis → garter snake

5. **Why is the toad considered an opportunistic feeder in this food web?**

   A. The toad is eaten by several different organisms.

   B. The toad is part of three different food chains.

   C. The toad's main predator is the hognose snake.

   D. The toad can survive under many different environmental conditions.

Questions 6–9 refer to the following illustrations:

A.

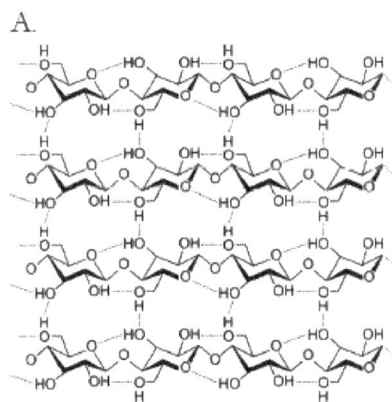

B.

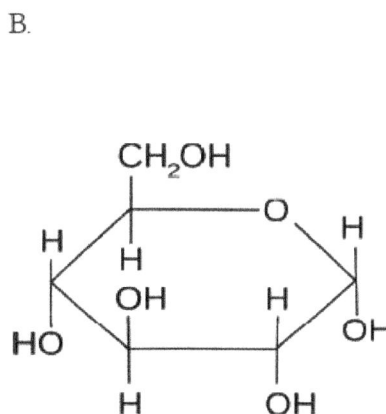

C.

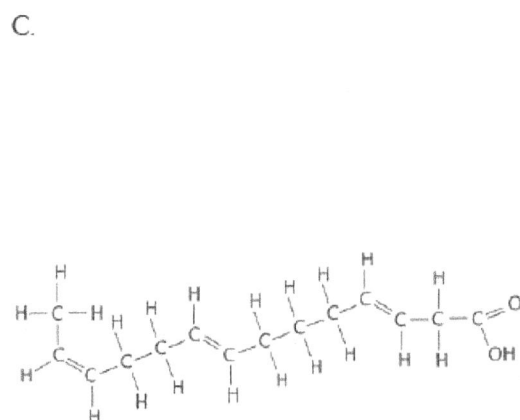

D.

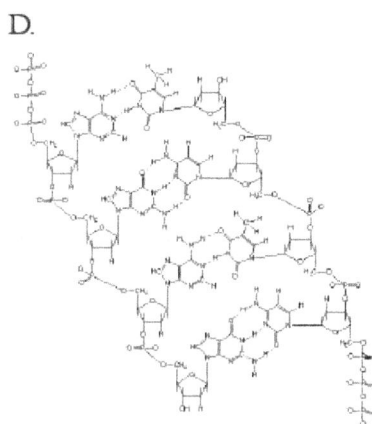

6. **Where is one most likely to find the structure labeled D?**

A. Inside the gall bladder

B. Within the cell nucleus

C. Outside the cell membrane.

D. Within the endoplasmic reticulum

7. **Molecule B is produced at the end of which biochemical process?**

A. ATP synthesis

B. Transcription

C. Photosynthesis

D. Cellular respiration

8. **What is the function of molecule C?**

   A. Long term energy storage

   B. Short term energy boost

   C. Bone construction and destruction

   D. Transmission of electrical impulses

9. **Structure A is a polysaccharide. It cannot be digested by most animals and is often found in the cell walls of plants. What is it?**

   A. Cellulose

   B. Starch

   C. Glycogen

   D. Sucrose

10. **Many chemical reactions that occur within living things produce energy. They can be represented A + B → AB + energy. What type of reaction is this?**

    A. Anabolic

    B. Dehydration

    C. Hydrolytic

    D. Exergonic

11. **The zebra mussel (Dreissena polymorpha) is a small bivalve that has been known to cause problems for boaters in the Great Lakes by attaching to their propellers and clogging up irrigation pipes, preventing water flow. These animals have few natural predators and are not native to the area. What is the term that describes this type of non-native organism?**

    A. Biodiversity

    B. Predator

    C. Invasive species

    D. Parasitic species

Chapter 4 Quiz Answer Key _____

1. **Answer: D.** To mimic something means to look a lot like it. The viceroy butterfly looks almost exactly the same as a monarch. Since it only takes one try for predators to realize that monarchs do not taste good, they are likely to remember the color pattern for a long time and avoid eating anything that bears that pattern.

2. **Answer: C.** In a food chain, there is a principle called the 10% rule. This means that only 10% of an organism's energy is available to the next trophic level, because the organism uses most of its energy to maintain its metabolism. Since there is so little energy available for the higher trophic levels, very few food chains can support more then five levels.

3. **Answer: A.** In this food web, garter snakes only eat grasshoppers. If the grasshoppers were to decrease in numbers, the garter snakes would have nothing to eat. Therefore, their population size would dramatically decrease.

4. **Answer: A.** The mouse occupies the tertiary consumer role in the food chain that starts with the grass, continues with the grasshopper, followed by the spider. The mouse then eats the spider.

5. **Answer: B.** In this food web, the toad is part of three different food chains. This means that if something happens to one of its food sources, it still has other opportunities to eat.

6. **Answer: B.** The molecule labeled D exhibits a double-helix structure, and therefore is DNA, which is found within a cell nucleus.

7. **Answer: C.** Molecule B is glucose. It is produced by autotrophs at the end of the Calvin Cycle in photosynthesis and is broken down by the organism for energy.

8. **Answer: A.** This molecule is a fatty acid. It is part of a lipid, which is used by the body to store energy for the long term. Short-term energy boosts are provided by carbohydrates because they are much easier than lipids to break down.

9. **Answer: A.** This huge molecule is cellulose, made from many repeating sugar molecules. It is found in the cell walls of plants and offers them support to stand upright toward the sun. Sucrose is a much smaller sugar. Glycogen is found in animal cells, and starch is another polysaccharide that is found in the roots of plants.

10. **Answer: D.** In exergonic reactions, heat is produced and will appear as one of the products in the equation. Hydrolytic reactions break things down by adding water into the system. Dehydration reactions have water as a product. Anabolic reactions take in heat.

11. **Answer: C.** The zebra mussel is an invasive species that arrived in the Great Lakes region most likely in ballast water from a cargo ship. Since it has no natural predators, it is able to survive and reproduce to the point where it has become disruptive.

SECTION III:
Practice Tests

Directions: For each question choose the best answer.

1. **Identify the variables in the following experiment: An investigation was done with chickens of different sizes. The number of eggs each chicken laid in a week was recorded.**

   A. Dependent variable was the number of eggs and the independent variable was the size of the chickens.

   B. Independent variable was the number of eggs and the dependent variable was the size of the chickens.

   C. Control variable was the number of eggs and the independent variable was the size of the chickens.

   D. Dependent variable was the number of eggs and the control variable was the size of the chickens.

2. **Monocots typically have parallel veins and flowers with petals in multiples of 3, whereas dicots typically have branched veins and flowers with petals in multiples of 4 or 5. Magnolia trees, however, have branched veins and flowers with petals in multiples of 3. What does this say about magnolias?**

   A. Magnolias are dicots, because they look like dicots. They do not look like grasses, irises, lilies, and other monocots.

   B. Dicots are not a monophyletic group, and magnolias are in a group of dicots different from other common dicots such as roses, mallows, and sunflowers.

   C. Monocots are a monophyletic group and therefore are the basal group, including grasses like wheat and corn but also trees like magnolias and date palms.

   D. Dicots are a diphyletic group composed of one branch with petals in multiples of 4-5 and another branch with atypical flowers such as magnolias and bananas.

3. **In a data set, what is the term for the value that occurs with the greatest frequency?**

   A. Mean

   B. Median

   C. Mode

   D. Range

4. Which of the follow statements is NOT true for both of the organelles shown?

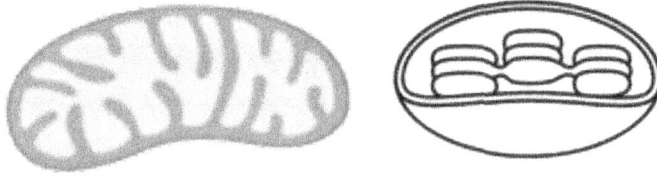

A. The inner membrane is the primary site for its activity.

B. They convert energy from one form to another.

C. They use an electron transport chain.

D. They are an important part of the carbon cycle.

Questions 5 – 7 refer to the information below.

A large population of laboratory animals has been allowed to breed randomly for a number of generations. After several generations, 49 percent of the animals display a recessive trait (bb), the same percentage as at the beginning of the breeding program. The rest of the animals show the dominant phenotype, with heterozygotes indistinguishable from the homozygous dominants.

5. What is the most reasonable conclusion that can be drawn from the fact that the frequency of allele b has not changed over time?

A. The population is undergoing genetic drift.

B. The two phenotypes are equally adaptive under laboratory conditions.

C. There has been a high rate of mutation of allele B to allele b.

D. There has been sexual selection favoring allele b.

6. What is the frequency of allele b in the gene pool?

A. 0.07

B. 0.49

C. 0.51

D. 0.70

7. **What proportion of the population is heterozygous (Bb) for this trait?**

   A. 0.09

   B. 0.34

   C. 0.42

   D. 0.51

8. **How do manufacturers make vegetable oils solid or semisolid at room temperature?**

   A. Adding hydrogen atoms to the double bonds in the fatty acid hydrocarbon chains

   B. Removing hydrogen atoms and forming additional single bonds in the fatty acid hydrocarbon chains

   C. Removing hydrogen atoms and forming additional double bonds in the fatty acid hydrocarbon chains

   D. Adding hydrogen atoms to the single bonds of the fatty acid hydrocarbon chains

9. **If a segment of DNA has a template strand of 5'-TAC GAT TAG-3', what will be the RNA that results from the transcription of this segment?**

   A. 3'-TAC GAT TAU-5'

   B. 3'-UAC GAU UAG-5'

   C. 3'-AUG CUA AUC-5'

   D. 3'-ATG CTA ATA-5'

10. **In mammals, nerve cells have spaces between them, across which signals must be transferred. How do these signals get across this gap?**

    A. Neural impulses cause the release of chemicals that diffuse across the gap.

    B. Sodium and potassium rapidly flux back and forth to carry the signals across the gap.

    C. Electrical currents of varying voltages are emitted from one side of the gap to another.

    D. The calcium within the axons and dendrites of nerves adjacent to a gap acts as the messenger.

11. **Why would double fertilization be a useful adaptation in plants?**

   A. Two sperm fertilizing one egg would provide greater genetic diversity in the offspring, providing a mosaic cellular effect, where different cells of the organisms have different genotypes.

   B. Fertilization of a plant by gametes from two separate plants would provide greater genetic diversity in the offspring and reduce the chance for heterozygous dominant alleles causing dysfunctions.

   C. Two sperm enter the plant embryo sac; one sperm fertilizes the egg, the other forms the endosperm, which will act as food for the developing embryo.

   D. Two sperm enter the animal embryo sac; one sperm fertilizes the egg, the other forms the yolk, which will act as food for the developing embryo.

12. **The following pedigree illustrates the pattern for an autosomal (non-sex-linked) gene that displays complete dominance. This pedigree only shows affected (shaded) and non-affected individuals.**

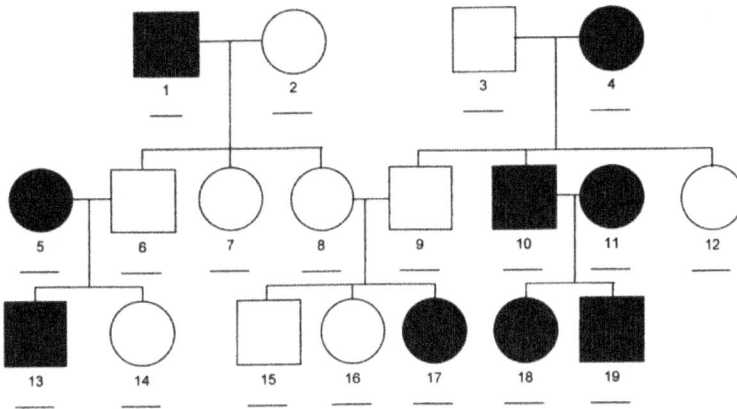

   Which members of the above pedigree could be homozygous dominant?

   A. 2 and 3

   B. 2, 9, and 12

   C. 6, 7, and 8

   D. 2, 15, and 16

13. During cellular respiration, there is an enormous transfer of electrons during redox reactions in the mitochondria. What is the result when hydrogen ions are pumped out of the mitochondrial matrix, across the inner mitochondrial membrane, and into the space between the inner and outer membranes?

A. Damage to the mitochondrion

B. The creation of a proton gradient

C. The lowering of pH in the mitochondrial matrix

D. The restoration of the Na-K balance across the membrane

14. The relative location of four genes on a chromosome can be mapped from the following data on crossover frequencies:

Gene	Frequency of Crossover
B and D	5%
C and D	50%
C and B	45%
A and B	30%
C and A	15%

Which of the following represents the relative positions of these four genes on the chromosome?

A. ABCD

B. BDCA

C. CABD

D. DBCA

15. A scientist investigating the mating behaviors of the zebra finch placed some of its eggs into the nest of another species of finch. They then allowed the zebra finches to be raised by the other species. Results showed that when the zebra finches were ready to start courting females, they turned their attention towards the species of the adoptive parents instead of their own species. Why do the zebra finches behave in this manner?

A. The zebra finches are acting out of habit.

B. The zebra finches have imprinted on their new parents.

C. The other species offers reinforcement techniques to select a mate.

D. The parents of the other species have conditioned the zebra finches.

16. **What is required when amino acid molecules bond together to form larger molecules?**

   A. The release of water molecules

   B. The increase of activation energy

   C. The addition of water molecules

   D. The release of carbon dioxide molecules

17. **What would a molecule formed from a hydrogen atom (which exhibits weak electronegativity) and an oxygen atom (which has relatively strong electronegativity) demonstrate?**

   A. Positive and negative partial charges on the oxygen and hydrogen atoms, respectively.

   B. Negative partial charges on both the oxygen and hydrogen atoms.

   C. No overall polarity.

   D. Negative and positive partial charges on the oxygen and hydrogen atoms, respectively.

18. **The T4 bacteriophage has a double-stranded DNA genome. How might infection with such a virus cause evolutionary change in a bacterial cell?**

   A. During replication, a progeny virus incorporates a gene from the infected bacterium into its capsid. It then infects another bacterium and introduces the DNA from the previously infected bacterium into the newly infected bacterium. The newly introduced DNA has no homologous gene on the main chromosome, so it incorporates into the chromosome as an insertion.

   B. After infection, the phage genome undergoes reverse transcription. The reverse-transcribed genes of the virus are incorporated into the bacterium's genome.

   C. During viral replication, mRNA copies of the infected bacterium's genes are incorporated into progeny virus. When the progeny viruses infect a new bacterium, these genes undergo reverse transcription and are integrated into the newly infected bacterium's genome.

   D. During viral replication, a progeny virus incorporates genes from an infected bacterium into its capsid. The progeny virus then introduces the DNA from the previously infected bacterium into a newly infected bacterium and this introduced DNA undergoes homologous recombination with the genome of this newly infected bacterium.

19. The fossil record shows that large flying insects arose approximately 250 million years ago. Evidence also suggests that the oxygen concentration of the atmosphere at this time was approximately 28% higher compared to present day levels. Knowing this, what physiological features allows mammals to attain great size, whereas today's insects remain relatively small?

    A. Red blood cells and the mammalian immune system.

    B. White blood cells and the mammalian immune system.

    C. Lungs and the mammalian circulatory system.

    D. The heart and the mammalian central nervous system.

20. A retrovirus infects a cell. Retroviruses have RNA genomes. How can this infection cause evolutionary change in the species?

    A. mRNA is transcribed from the retroviral genome and this mRNA is reverse transcribed and then incorporated into the host genome using integrase.

    B. Genes from the retroviral genome are directly incorporated into the host cell genome using homologous recombination.

    C. Host mRNA and genes from the retroviral genome are reverse transcribed and then incorporated into the host genome using integrase.

    D. Retroviral proteins are converted into DNA by protease, and then integrated into the host genome using integrase.

21. cDNA is an mRNA sequence that has been reverse transcribed into DNA. Which of the following could be true?

    A. cDNA can be created by a virus; it will be an exact copy of the gene from which the mRNA was transcribed, causing a point mutation.

    B. cDNA can be created by the use of reverse transcriptase by molecular biologists; it will be larger than the gene from which the mRNA was transcribed.

    C. cDNA can be created by either a virus or a molecular biologist using reverse transcriptase; it will be an exact copy of the gene from which the mRNA was transcribed.

    D. cDNA can be created by either a virus or a molecular biologist using reverse transcriptase; it will be smaller than the gene from which the mRNA was transcribed.

22. **The worldwide population of the North Atlantic right whale has dramatically decreased, almost to the point of extinction. It also shows very little genetic diversity. Which of these is the most important risk factor the whales face as a result of this?**

    A. The habitat of the whales is threatened by climate change and global warming.

    B. A dwindling food supply makes it more difficult for the right whales to survive and reproduce.

    C. Mutations are more likely to affect more individuals in populations with low genetic diversity.

    D. The low genetic diversity interferes with the ability of the population to respond to environmental changes.

23. **An antiviral drug inhibits the protein reverse transcriptase. Against what type of virus will the drug be effective?**

    A. Bacteriophage T4, a lytic phage with a double-stranded DNA genome

    B. HIV, a lentivirus with a double-stranded DNA genome

    C. Parvovirus B19, a parvovirus with a single-stranded DNA genome

    D. SIV, a lentivirus with a single-stranded RNA genome

24. **Algae, including seaweed, and plants are similar in that they are both eukaryotic organisms that create their energy from sunlight through photosynthesis. Which of the following are fundamental differences between algae and plants?**

    A. The evolution of xylem in plants has allowed them to become much larger than algae.

    B. Flowers allow plants to be fertilized by insects; this kind of symbiotic relationship is not seen with algae.

    C. In the ocean, swimming gametes can meet, as sperm cells meet egg cells. On land, seeds are needed to allow fertilization to occur.

    D. Land plants require tissue differentiation that is not necessary in algae.

25. **A scientist wants to find fossils from dinosaurs that lived at the end of the Jurassic period. Where would be a reasonable place to look?**

    A. Igneous rocks from the late Triassic period (the era just preceding the Jurassic)

    B. Sedimentary rocks from the late Triassic period (the era just preceding the Jurassic)

    C. Igneous rocks from the early Cretaceous period (the era just after the Jurassic)

    D. Sedimentary rocks from the early Cretaceous period (the era just after the Jurassic)

26. **Myoglobin is located in the muscles. Which of the following are true about myoglobin?**

   A. Myoglobin has a higher affinity for oxygen than hemoglobin because muscles need to receive oxygen from blood.

   B. Hemoglobin has a higher affinity for oxygen than myoglobin because of the importance of transporting oxygen throughout the body.

   C. Hemoglobin has a higher affinity for oxygen than myoglobin because it needs to also carry CO_2, a waste product of respiration.

   D. Myoglobin has a higher affinity for oxygen than hemoglobin because muscles sometimes need to undergo anaerobic respiration.

27. **Which of the following could be expected to lead to anemia in a mammal, and why?**

   A. A deficiency of iron, because iron is important for functional heme

   B. A deficiency of iron, because iron carries oxygen into the nucleus of red blood cells

   C. Genetic disorders that cause abnormal hemoglobin, because this will result in a lower number of red blood cells

   D. Genetic disorder that cause a higher-than-normal number of red blood cells, but without any functional hemoglobin

28. **In mosses, the dominant part of the plant is the gametophyte, the velvety green part that looks a bit like a miniature lawn. Gametes are formed from this part by mitosis. What is true about this velvety, dominant stage of the moss?**

   A. It is haploid; this differentiates mosses from vascular plants, in which the sporophyte is dominant.

   B. It is diploid; this differentiates mosses from vascular plants, in which the sporophyte is dominant.

   C. It is haploid, as is the sporophyte in vascular plants.

   D. It is diploid, as is the gametophyte in vascular plants.

29. **The lack of a nucleus in many vertebrate red blood cells has been proposed as a reason for the accumulation of "junk" or vestigial non-coding DNA in those genomes. What evolutionary factors support this view?**

    A. The nucleus helps to control the amount of non-coding DNA; without nuclei in their red blood cells, these vertebrates are unable to "clean up" junk DNA.

    B. Not having nuclei in red blood cells saves energy that can be used in other cells to maintain junk DNA.

    C. In organisms with nuclei in their red blood cells, the cells are bigger, so in order to fit in small blood vessels, there is selection pressure to keep the nuclei smaller. In animals that do not have nuclei in their red blood cells, larger nuclei do not affect the size of the red blood cells, so this selection pressure is absent.

    D. In organisms with nuclei in their red blood cells, the efficiency of oxygen transport is greater, so there is greater selection pressure to retain their nuclei. This selection pressure is absent in animals that do not have nuclei in their red blood cells, allowing the accumulation of junk DNA.

30. **Research has demonstrated that some cancer patients with a poor prognosis have high levels of a specific receptor tyrosine kinase called HER2. Which of the following reasons would explain why creating a drug to bind HER2 might inhibit tumor growth?**

    A. Binding HER2 would increase the phosphorylation of downstream proteins.

    B. Binding HER2 would prevent the phosphorylation of downstream proteins.

    C. Binding HER2 would allow the flow of ions through the membrane.

    D. Binding HER2 would block the flow of ions through the membrane.

31. A flower is most commonly pink, but occasionally white or yellow versions are observed. After several crosses, the following data were obtained:

Parents	F1	F2
Pink x white	Pink	195 Pink, 64 white
Pink x yellow	Pink	150 Pink, 53 yellow
white x yellow	Pink	228 Pink, 97 white, 76 yellow

Which of the following statements best explains the data?

A. The presence of pink in all three groups of the F1 generation indicates that this phenotype is a function of the environmental conditions present during the F1 cross that are different from those during the F2 cross

B. Pink and yellow are inherited, segregated genes that follow Mendel's laws, but white is a function of the flowers not having matured enough for their color to show

C. Pink and yellow follow Mendel's laws but white demonstrates non-Mendelian inheritance; white probably results from vegetative (clonal) reproduction

D. Pink, white, and yellow are all inherited genes. The inconsistency could be due to a small sample size, or it could be due to some other factor, such as another gene influencing the phenotypic expression of the flowers.

32. **After doing several more crosses (for the same flowers from the previous question), the data look like this:**

Parents	F1	F2
Pink x white	Pink	740 Pink, 230 white
Pink x yellow	Pink	760 Pink, 270 yellow
white x yellow	Pink	480 Pink, 260 white, 240 yellow

Which of the following statements can now be made with certainty?

A. The traits are inherited, pink is dominant, and the data conform to Mendel's law of independent assortment.

B. The traits are inherited, pink and yellow are codominant, and the data conform to Mendel's law of independent assortment.

C. The traits are not inherited.

D. The traits are inherited, pink is dominant, and the data conform to Mendel's law of segregation.

33. A patient is diagnosed with a rare viral infection exhibiting symptoms including a severe cough that causes hemorrhaging of the lungs. The coughed up blood contains viable virus particles, but conventional PCR for DNA does not produce any visible bands on an electrophoresis gel. Why are the viral particles not likely to be detected with this method?

    A. A protein coat surrounding the nucleic acid prevents the DNA from amplifying.

    B. The virus is an RNA virus and would need reverse transcription to be detected by amplification.

    C. The surrounding a protein coat only contains the base plates and the tails of the virus.

    D. The DNA concentration is too low in the blood samples for PCR amplification.

34. The graph shows the results of an investigation into the amount of energy (in the form of heat) produced by different sized animals.

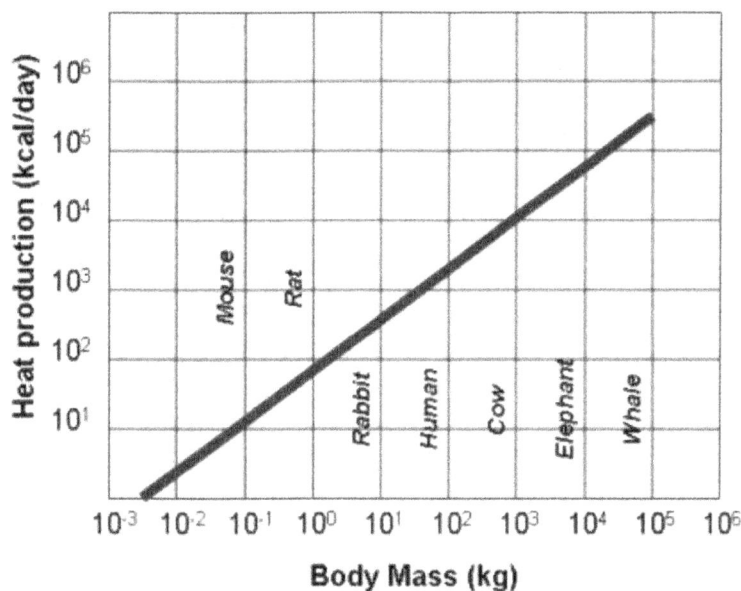

What conclusion can be made from the information provided in the graph?

    A. Animals that have a large body mass also produce large amounts of heat.

    B. Body mass is inversely related to the amount of heat produced by an animal.

    C. The individual cells of small animals produce less heat than those of large animals.

    D. Larger animals possess adaptations for dissipating heat more effectively than smaller animals.

35. Hyperventilation can cause the pH of the blood to rise. Considering this chemical equation:

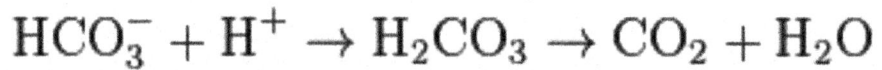

$$HCO_3^- + H^+ \rightarrow H_2CO_3 \rightarrow CO_2 + H_2O$$

hyperventilation will result in which of the following?

A. Respiratory acidosis, because CO_2 is released in a greater amount, driving this reaction to the right and making the blood more acidic.

B. Respiratory alkalosis, because CO_2 is released in a greater amount, driving this reaction to the right and making the blood more basic.

C. Respiratory acidosis, because CO_2 is released in a greater amount, driving this reaction to the left and increasing the concentration of hydrogen ions.

D. Respiratory alkalosis, because CO_2 is released in a greater amount, driving this reaction to the left and increasing the amount of H_2CO_3.

36. The graph shows the rates of survivorship for different organisms.

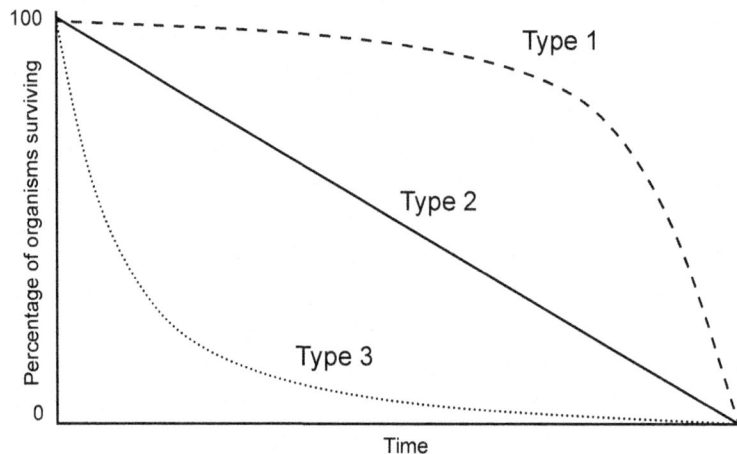

Which of the following pairs best matches the expected curves for field mice and humpback whale?

A. Field mouse – Type III; humpback whale – Type I

B. Field mouse – Type I, humpback whale – Type III

C. Field mouse – Type II; humpback whale – Type II

D. Field mouse – Type I; humpback whale – Type I

37. **Using the following taxonomic key, identify the tree from which this branch came.**

1 - Are the leaves PALMATELY COMPOUND (BLADES arranged like fingers on a hand)? – go to 2

1 - Are the leaves PINNATELY COMPOUND (BLADES arranged like the veins of a feather)? – go to 3

2 - Are there usually 7 BLADES - *Aesculus hippocastanum*
2 - Are there usually 5 BLADES - *Aesculus glabra*

3 - Are there mostly 3-5 BLADES that are LOBED or coarsely toothed? - *Acer negundo*
3 - Are there mostly 6-13 BLADES with smooth or toothed edges? - *Fraxinus americana*

   A. *Aesculus hippocastanum*

   B. *Aesculus glabra*

   C. *Acer negundo*

   D. *Fraxinus americana*

38. **It is thought that mitochondria evolved from symbiotic bacteria that were engulfed by a larger cell. Which of the following represents evidence that supports this?**

   A. Mitochondria are smaller than the eukaryotic cells; eukaryotic cells have linear chromosomes

   B. Mitochondria have their own DNA; eukaryotic genomes are phylogenetically similar to the archaea

   C. Mitochondria encode ATP synthase; eukaryotic cells encode the machinery for glycolysis

   D. Mitochondria have double membranes; eukaryotic genomes are phylogenetically similar to the eubacteria

39. Chloroplasts are thought to have evolved from endosymbiosis of photosynthetic cyanobacteria. Which of the following represents evidence that supports this?

   A. Chloroplasts conduct photosynthesis using H_2O and light; chloroplasts are found only in plants.

   B. Cyanobacteria conduct photosynthesis using H_2O and light; some free-living chloroplasts persist in remote environments.

   C. Chloroplasts and cyanobacteria both have thylakoid membranes; the genome sequences of chloroplasts are homologous to those of cyanobacteria.

   D. Cyanobacteria conduct photosynthesis using O_2 and light; chloroplasts have thylakoid membranes like cyanobacteria.

40. Yellow squash and zucchini plants sometimes spontaneously mutate to become toxic, even deadly. Plants evolve in this way to protect themselves from being eaten. What difference between plants and animals underlies this evolutionary pattern?

   A. Plants lack immune systems.

   B. Plants lack digestive systems.

   C. Plants lack nervous and musculoskeletal systems.

   D. Plants can mutate easier because they can reproduce clonally.

41. A woman has Pearson Syndrome, a disease caused by a mutation in mitochondrial DNA. In which of the following individuals would you expect to see the disease?

   I   Her daughter
   II   Her son
   III Her daughter's son
   IV   Her son's daughter

   A. I, III

   B. I, II, III

   C. II, IV

   D. I, II, III, IV

42. Photosynthetic organisms like plants and algae use H_2O as a reactant. However, some deep-sea organisms survive in the complete absence of light. Which two things could be true about such organisms?

A. They could produce their own light through bioluminescence or they could be heterotrophs

B. They could be heterotrophs or they could be algae that translocate between light and dark levels of the ocean

C. They could be oligotrophs that use sulfur from deep ocean vents as energy, or they could be heterotrophs

D. They could be heterotrophs or they could be autotrophs that use H_2S as a reactant.

43. Which of the following would be considered an example of apoptosis that occurs during the embryological development of an animal?

A. An increase in the growth rate of bone forming cells that increase bone length in the animal.

B. The death and disintegration of certain cells in the developing limb to form fingers and toes.

C. A decrease in the rate of cell division of liver tissue to ensure development does not occur too quickly.

D. The activation of regulatory genes in the cells that control the contraction and relaxation of the the developing heart.

44. The archaea are prokaryotic organisms with genomes that are very different from the genomes of eubacteria, and more similar to eukaryotic genomes. They tend to be found in extreme environments, such as extremely salty (halophiles), extremely hot (thermophiles), and extremely low pH (acidophiles). What might explain this phenomenon?

A. Archaea were originally ubiquitous, but when some archaean ancestors formed symbioses with eubacteria to become eukaryotes, they outcompeted the archaea. Only those archaea who could handle extreme conditions survived, and the traits that allowed this were selected by natural selection.

B. Archaea have always been extremophiles, since life was created. The original world contained archaea, eubacteria, and eukaryotes, but the eukaryotes expanded because of their superior survival capabilities; this is an example of natural selection after creation.

C. All original life forms were eubacteria, and archaea evolved from them in extreme environments. Evidence for this is that they are both prokaryotes; eukaryotes have a nucleus and organelles, cell structures that evolved independently from eubacteria in mild conditions at a later date.

D. Eubacteria were originally ubiquitous, but then some formed endosymbiotic relationships, engulfing archaea to become eukaryotic organisms capable of respiration. The archaea that lived in extreme environments, however, were not in contact with eubacteria and escaped the fate of becoming organelles.

45. Mitochondrial genomes are much smaller than nuclear genomes, and most of the proteins used by mitochondria are encoded in the nucleus and subsequently transported to the mitochondria. Many of these nuclear genes are phylogenetically similar to genes from bacteria. Why might this be?

    A. The nucleus of the original eukaryotic cell encoded proteins needed by the mitochondrion; this is why the evolutionary event of endosymbiosis occurred.

    B. At the time of the ancient endosymbiosis event, the mitochondria encoded all of their genes, but some genes were transferred from the mitochondria to the nucleus through lateral gene transfer.

    C. At the time of the ancient endosymbiosis event, the mitochondria encoded all of their genes, but then, the mitochondria began using nuclear genes, and the mitochondrial genes were eliminated from the mitochondria to avoid redundancy.

    D. The nuclear genome evolved superior genes that accomplished the functions of the original mitochondrial genes, but with greater efficiency. This is an example of convergent evolution.

46. The diagram represents two nerve cells that are adjacent to each other. Which sequence of events correctly describes the transmission of the nerve impulse as it travels from point A to point E?

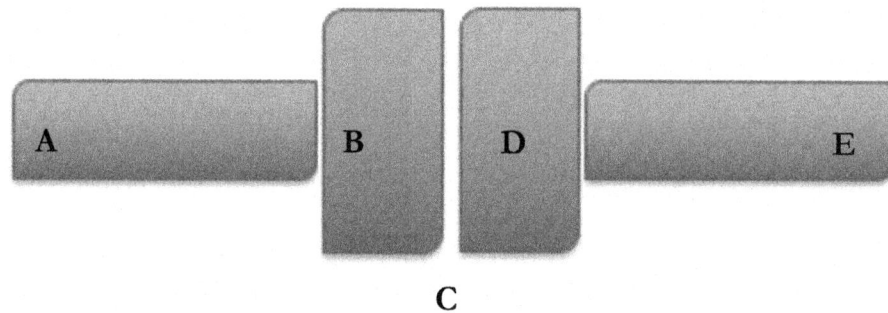

C

    A. At point A, an influx of Na and outflow of K initiates the action potential, which travels to the end of the axon (B). Neurotransmitters are released into the synapse (C) and are absorbed by the next axon (D). The neurotransmitter initiates an action potential, which travels to point E.

    B. At point A, an influx of Na and outflow of K initiates the action potential, which travels to point B. The synapse (C) is flooded with Na that is absorbed by the next axon (D), thus continuing the action potential to point E.

    C. At point A, an influx of neurotransmitters initiates the action potential, which travels to point B. Na is released into the synapse (C) and is absorbed by the next axon (D). The Na stimulates the influx of neurotransmitters in the next axon and the action potential travels to point E.

    D. At point A, an influx of Na stimulates the release of neurotransmitter within the axon cell. The neurotransmitters flow to the end of the axon (B), cross the synapse (C), and continue to point E.

47. A plant was grown in a test tube containing radioactive nucleotides, the molecules from which DNA is built. When examining the dividing cells in the plant after several days of growth, where would the researchers expect the majority of the radioactivity to be concentrated?

A. Rough endoplasmic reticulum

B. Smooth endoplasmic reticulum

C. Central vacuole

D. Nucleus

48. A man and a woman both claim to be distant descendants of a Saudi Arabian king who lived over a century ago. A DNA test reveals that the man is definitely his descendent, but no such statement can be made for the woman. Why can paternity be more definitively proven with the son than with the daughter?

A. Since women have two X chromosomes, the DNA from one interferes with the DNA from the other, preventing sequencing.

B. Only men have a Y chromosome, and this is a smaller chromosome that encodes fewer genes, allowing less ambiguity when comparing the sequences.

C. Only men have a Y chromosome, precluding recombination, so a descendant through the male line will have an almost exact copy of the Y chromosome sequence of his ancestor.

D. While the sequence of one X chromosome of a woman comes from her father, the genes are inactivated though X-inactivation and this epigenetic modification prevents a definitive comparison.

49. Some studies have suggested that smallpox is more severe in people with type A blood than people with type O or type B blood. Why might this be?

A. Type A blood cells have membranes that are more vulnerable to viral entry than those of other blood types

B. The smallpox virus encodes anti-type A antibodies that attack the blood of type A humans.

C. The antigen on the nuclear membrane of type A blood cells binds the smallpox virus, facilitating viral entry.

D. The smallpox virus has surface antigens that resemble type A blood, allowing it to "hide" in humans with type A blood.

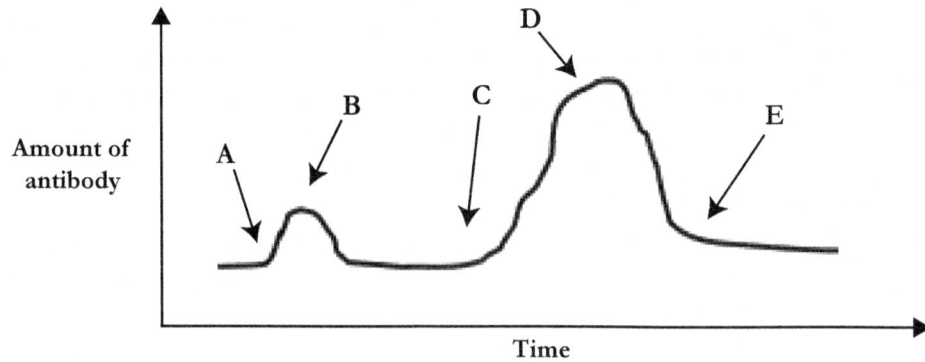

50. **Which of the points above might represent a vaccination and a booster shot?**

A. A and C

B. B and D

C. A and B

D. C and D

51. **What might point C in the graph represent?**

A. Exposure to a viral infection after vaccination against the virus

B. Exposure to a viral infection after a previous exposure to the virus

C. Both A and B

D. Neither A nor B; antibodies are only produced in response to bacteria

52. **If the graph represents response to a viral infection, why is point D higher than point B?**

A. The antibodies at point B have been supplemented by additional antibodies at point D; point D represents the sum of the antibodies produced in response to the antigen at point A and the antigen at point C.

B. The infecting virus at point C was stronger; for example, cowpox and smallpox both produce the same kinds of antibodies, but smallpox induces more antibodies than cowpox.

C. The person ingested zinc and vitamin C, known immune boosters, at point C, which induced the production of antibodies.

D. The antibodies produced at point B were produced and then levels returned to normal; however, the immune system was primed by the exposure at point A to be capable of efficiently producing a greater number of antibodies in response to the antigen at point C.

53. **What might the graph look like if the virus were to once again infect this person shortly after point E?**

   A. A peak similar in size to point D would be observed. The same B cells that recognized the antigen at point C would recognize a subsequent exposure to the antigen.

   B. A peak twice as large as that at point D would be observed; every exposure will double the immune response since more antibodies are added each time.

   C. A smaller peak than that seen at point D would be observed, because the immune system would have become habituated to the antigen.

   D. No peak would be observed, because the antibodies would be used up and would need time to build up again.

54. **Babies receive antibodies through their mother's milk. This is known as passive immunity. If a baby is exposed to an antigen and the mother's antibodies fight off the infection, what might a subsequent infection look like after the baby has been weaned?**

   A. It will look the same as a vaccination; whether the baby fought off the infection through the mother's antibodies or its own antibodies after vaccination makes no difference in the response.

   B. It will be a much smaller response than if the baby had been vaccinated. This is because a vaccination induces the production of memory cells in the baby whereas passive immunity does not.

   C. It will be much larger than if the baby had been vaccinated because the passive immunity produces a population of memory cells and the memory cell population will increase with exposure to the infection.

   D. There will be no response. Passive immunity causes a baby's immune system to become dependent on the mother's antibodies, so that after weaning, it is too weak to produce its own.

55. **In humans, the hormone testosterone enters cells and binds to specific proteins, which in turn bind to specific sites on the DNA of the cell. What is the function of these proteins?**

   A. Help RNA polymerase transcribe certain genes

   B. Promote recombination

   C. Unwind the DNA so that its genes can be transcribed

   D. Cause mutations in the DNA

56. **Based on the pedigree chart below, what term best describes the nature of the trait being mapped?**

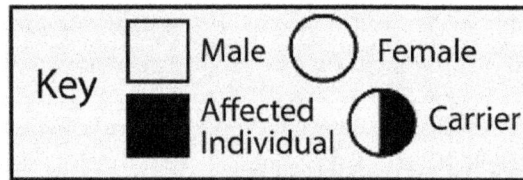

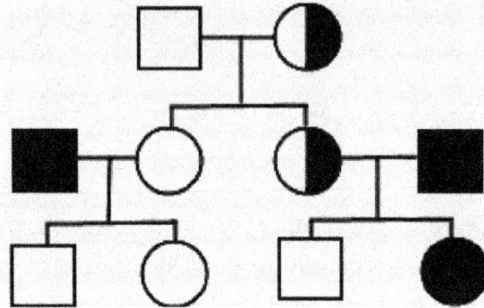

A. Autosomal recessive

B. Sex-linked

C. Incomplete dominance

D. Co-dominance

57. **Capillaries come into contact with a large surface of both the kidneys and lungs, especially in relation to the volume of these organs. Which of the following is not consistent with both organs and their contact with capillaries.**

A. Small specialized sections of each organ contact capillaries

B. A large branching system of tubes within the organ

C. A large source of blood that is quickly divided into capillaries

D. A sac that contains a capillary network within its interior volume

58. Samples of DNA were taken from a common species of tree frog in the temperate zones of North America. Samples came from individual frogs across the entirety of their range; however, upon analysis, researchers observed low genetic diversity in an isolated population found in the Ohio Valley. Based upon this observation, what would you hypothesize happened in the recent history of this population?

A. A population bottleneck occurred with little to no migration into the valley.

B. Habitat loss and climate change introduced extrinsic forces that are selecting upon the population to adapt to a specialized niche.

C. Adaptive radiation for living in the Ohio Valley has resulted in this population.

D. The population experiences lower levels of genetic drift as a result of the small population.

59. The term "vaccination" is etymologically related to the Latin word for cow, "vacca". Why?

A. The first species to be vaccinated was the cow, an animal of great economic importance in the 19th century. Veterinarians used an inactivated version of cowpox to protect cows against this deadly bovine disease.

B. Scientists originally investigated vaccines in the cow. Animal experiments in cows indicated that vaccines could be safely and effectively used in humans.

C. Cowpox is a similar virus to the serious human virus, smallpox, and was used as the original vaccine because it causes a mild infection in humans. The idea for this came from the observation that milkmaids did not contract smallpox.

D. When polio infects cows, it mutates into an attenuated form. When this attenuated virus is injected into humans, it causes a mild infection that serves as a vaccination against polio.

60. In an aqueous solution, in which way do most of the hydrophobic side chains of amino acids orient themselves?

A. Exposed to the exterior of the protein to promote interactions with other molecules.

B. In a configuration that promotes their interaction with the peptide backbone.

C. Pointed inward such that they associate with one another.

D. In a random manner determined by the positioning of hydrophilic side chains.

61. **Which of these would result from incapacitating the muscles of a mammalian eye?**

    A. The inability to regulate the amount of light entering the eye and an inability of the photoreceptor cells to transduce light energy

    B. The inability of the photoreceptor cells to transduce light energy

    C. The inability to focus light and an inability of the photoreceptor cells to transduce light energy

    D. The inability to regulate the amount of light entering the eye and an inability to focus light

62. **A sequence of pictures of polypeptide synthesis shows a ribosome holding two transfer RNAs. One tRNA has a polypeptide chain attached to it; the other tRNA has a single amino acid attached to it. What does the next picture show?**

    A. Both tRNAs remain, but the polypeptide chain is transferred to the single amino acid.

    B. The tRNA connected to the single amino acid leaves the ribosome.

    C. The tRNA connected to the polypeptide chain leaves the ribosome.

    D. Both tRNAs remain, and a third tRNA with an amino acid joins the pair on the ribosome.

63. **The chemical reaction for photosynthesis is:**

$$6 CO_2 + 12 H_2O + \text{light energy} \rightarrow C_6H_{12}O_6 + 6O_2 + 6H_2O$$

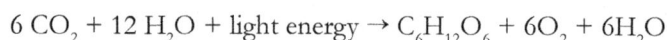

   A scientist used "heavy" water, with an oxygen with 10 neutrons instead of the normal 8 (^{18}O), as the water reactant, and finds that the O_2 gas released is composed of 100% ^{18}O. From these data, which of the following likely occurs during the reaction?

    A. The CO_2 is split into 6 carbon atoms and 6 O_2 gas molecules, using NADPH

    B. The CO_2 is split into 6 carbon atoms and 6 O_2 gas molecules, using light energy

    C. The oxygen in the water molecules in the product come from the CO_2 molecules

    D. The 12 H_2O molecules are split, and 6 oxygen atoms are incorporated into $C_6H_{12}O_6$ and 6 oxygen atoms are incorporated into H_2O

Part II. Grid-in Questions.

1. A plant geneticist is investigating the inheritance of genes for sour taste (Su) and green skin (e) in apples (*Malus domestica*). Green skin is recessive and the sour taste are a result of the recessive genotype (susu). The geneticist wishes to determine if the genes assort independently and performs a testcross between a sour/red hybrid and a tree homozygous recessive for both traits. The following offspring are produced:

   sour/red – 48
   sour/green – 28
   non-sour/red – 22
   non-sour/green – 41

   Calculate the chi-squared value for the null hypothesis that the two genes assort independently. Give your answer to the nearest tenth.

2. In fruit flies, Drosophila, a phenotype of eyes that look like small slits or kidney shaped "bar eye" is a result of a mutation displaying an autosomal dominant pattern of inheritance. Homozygous recessive individuals (B^+B^+) display a wild-type phenotype with normal eyes. Inheriting two copies of the mutation (BB) is lethal during embryonic development. In a cross between strains of drosophila with the bar eye phenotype, what proportion of eggs laid is expected to be bar eyed? Give your answer in the form of a fraction.

3. The following are lengths of bull head minnows, measured from nose to tail, recorded in centimeters. Calculate the mean. Give the answer in cm to the nearest tenth.

   | 15.4 | 10.2 | 14.6 | 9.1 | 9.2 | 13.1 | 12.5 | 15.2 | 11.2 |

4. Calculate the standard deviation for the data set given in the previous question. Give the answer in cm to the nearest tenth.

5. In peas (*Pisum sativum*), wrinkled peas (W) are dominant to smooth peas (w). Peas from the offspring of a cross between wrinkled-pea and smooth-pea plants were used in a lab. A student counts 429 wrinkled and 399 smooth peas in the pod. Calculate the chi-squared value for the null hypothesis that the purple parent was heterozygous for wrinkled peas. Give your answer to the nearest tenth.

Chi-Square Table								
Degrees of Freedom								
p	1	2	3	4	5	6	7	8
0.05	3.84	5.99	7.82	9.49	11.07	12.59	14.07	15.51
0.01	6.64	9.32	11.34	13.28	15.09	16.81	18.48	20.09

6. A tropical plant displays red flower color or pink flower color, with red the dominant phenotype, in an autosomal dominant pattern of inheritance. In a cross between two heterozygous strains of the flower, what proportion of offspring plants will have pink flowers? Give your answer in the form of a fraction.

Part III. Free Response Questions. _____

There are eight questions to answer in this section. The first six are short response. These will require a paragraph or two of explanation and discussion. Consider spending about 6 minutes on each one of these. The last two items require a longer response. Read each part of the item carefully and spend about 20 minutes answering each one.

1. Describe the process of signal transduction, including an example.

2. The Endosymbiotic Theory proposes that mitochondria and chloroplasts originate through symbiosis. <u>Describe</u> the characteristics of these organelles that led scientists to hypothesize that they were once free-living organisms.

3. Name three differences between plants and fungi.

4. What are the three steps involved in the structural changes that occur during the mitotic process?

5. Explain how stabilizing selection works in an ecosystem. Provide an example.

6. Describe the process of ecological succession after a volcanic explosion.

7. When plants moved onto land, they needed to evolve several structures that allowed them to survive in an environment with much less water. These adaptations have afforded plants much success on land.

   (A) Briefly **discuss** the evolution of land plants from their aquatic ancestor. Include from where plants are believed to have evolved.

   (B) **Describe** three structural adaptations and functions that plants evolved in order to live on land.

   (C) **Explain** the relationship that evolved between plants and nitrogen-fixing symbiotic bacteria in the soil.

8. Describe the materials or factors involved in the photosynthetic reaction.

Question Number	Correct Answer	Your Answer	Question Number	Correct Answer	Your Answer
1	A		33	B	
2	B		34	A	
3	C		35	B	
4	A		36	A	
5	B		37	C	
6	D		38	B	
7	B		39	C	
8	A		40	C	
9	C		41	B	
10	A		42	D	
11	C		43	B	
12	D		44	A	
13	B		45	B	
14	C		46	A	
15	B		47	D	
16	A		48	C	
17	D		49	D	
18	D		50	A	
19	C		51	C	
20	C		52	D	
21	D		53	A	
22	D		54	B	
23	D		55	A	
24	D		56	B	
25	D		57	D	
26	A		58	A	
27	A		59	C	
28	A		60	C	
29	C		61	D	
30	B		62	C	
31	D		63	C	
32	D				

Practice Test I Answer Key _____

1. **Identify the variables in the following experiment: An investigation was done with chickens of different sizes. The number of eggs each chicken laid in a week was recorded.**

   A. Dependent variable was the number of eggs and the independent variable was the size of the chickens.

   B. Independent variable was the number of eggs and the dependent variable was the size of the chickens.

   C. Control variable was the number of eggs and the independent variable was the size of the chickens.

   D. Dependent variable was the number of eggs and the control variable was the size of the chickens.

   Answer: A.

   Dependent variable was the number of eggs and the independent variable was the size of the chickens. The dependent variable was measured by the investigators, number of eggs in a week. The independent variable was the difference between the chickens being tested, in this case size.

2. **Monocots typically have parallel veins and flowers with petals in multiples of 3, whereas dicots typically have branched veins and flowers with petals in multiples of 4 or 5. Magnolia trees, however, have branched veins and flowers with petals in multiples of 3. What does this say about magnolias?**

   A. Magnolias are dicots, because they look like dicots. They do not look like grasses, irises, lilies, and other monocots.

   B. Dicots are not a monophyletic group, and magnolias are in a group of dicots different from other common dicots such as roses, mallows, and sunflowers.

   C. Monocots are a monophyletic group and therefore are the basal group, including grasses like wheat and corn but also trees like magnolias and date palms.

   D. Dicots are a diphyletic group composed of one branch with petals in multiples of 4-5 and another branch with atypical flowers such as magnolias and bananas.

   Answer: B.

   Dicots are not a monophyletic group, and magnolias are in a group of dicots different from other common dicots such as roses, mallows, and sunflowers. While monocots are a true monophyletic group, dicots are not. From the basal angiosperm ancestor, the magnolia branch diverged before the monocot branch. The eudicot branch diverged later still.

3. **In a data set, what is the term for the value that occurs with the greatest frequency?**

A. Mean

B. Median

C. Mode

D. Range

Answer: C.

Mode. Mean is the mathematical average of all the items. The median depends on whether the number of items is odd or even. If the number is odd, then the median is the value of the item in the middle. Mode is the value of the item that occurs the most often, if there are not many items. Bimodal is a situation where there are two items with equal frequency. Range is the difference between the maximum and minimum values.

4. **Which of the follow statements is NOT true for both of the organelles shown?**

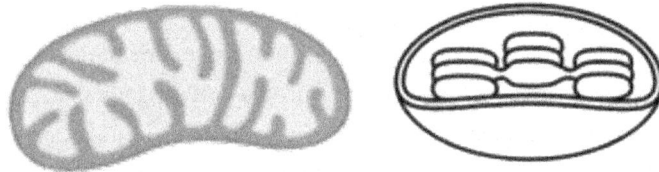

A. The inner membrane is the primary site for its activity.

B. They convert energy from one form to another.

C. They use an electron transport chain.

D. They are an important part of the carbon cycle.

Answer: A.

The inner membrane is the primary site for its activity. In mitochondria the electron transport chain is present in the inner membrane, however in chloroplasts it is present in the thylakoid membranes. Unlike mitochondria, which have two membranes, chloroplasts have three, an outer membrane, an inner membrane, and the thylakoid membrane, which is the innermost membrane in the chloroplasts. The inner membrane in a chloroplast, in other words, is really the middle membrane.

Questions 5 – 7 refer to the information below.

A large population of laboratory animals has been allowed to breed randomly for a number of generations. After several generations, 49 percent of the animals display a recessive trait (bb), the same percentage as at the beginning of the breeding program. The rest of the animals show the dominant phenotype, with heterozygotes indistinguishable from the homozygous dominants.

5. What is the most reasonable conclusion that can be drawn from the fact that the frequency of allele b has not changed over time?

A. The population is undergoing genetic drift.

B. The two phenotypes are equally adaptive under laboratory conditions.

C. There has been a high rate of mutation of allele B to allele b.

D. There has been sexual selection favoring allele b.

Answer: B.

Recessive traits will remain in a population if they are not harmful. There is nothing inherent about recessive traits that will make them rare; only that they will not be expressed in heterozygotes. Therefore, if the traits are equally adaptive, we would not expect to see any change in the allele frequencies. Now, if the situation is the interbreeding of two homozygous populations, the F1 generation will all have the dominant phenotype. This change in phenotype frequency is temporary, though—the next generation will see the re-emergence of the recessive phenotype, and if it is equally adaptive, its frequency will remain the same in subsequent generations. The alleles will be shuffled around but will not differ in their overall frequency in the population.

6. **What is the frequency of allele b in the gene pool?**

    A. 0.07

    B. 0.49

    C. 0.51

    D. 0.70

Answer: D.

0.70. If 49% of the population is bb, Using the Hardy-Weinberg equation: $1 = p^2 + 2pq + q^2$ If 49% of the population is bb, then $q^2 = .49$. From this, $q = .7$, which makes $p = .3$. The frequency of b is 70% and the frequency of B is 30%.

It is always a good idea to confirm these types of answers. Plugging back into the equation, if $q = .70$ and $p = .30$, then:

$p^2 + 2pq + q^2$

$= (.30)^2 + 2(.30)(.70) + (.70)^2$

$= .09 + .42 + .49 = 1.$

If you have time, checking your work using this equation can make sure you choose the correct answer. If you are short on time during the test, you will still be able to get the answer by taking the square root of the frequency of the homozygous recessive trait. If you are very short on time, you could quickly eliminate answers A and B, since you know that the frequency of allele b must be higher than 49% since that is the frequency of bb.

7. **What proportion of the population is heterozygous (Bb) for this trait?**

    A. 0.09

    B. 0.34

    C. 0.42

    D. 0.51

Answer: B.

0.34 .If 49% of the population is bb, then 51% of the population is a mix of BB, Bb, and bB, 1/3 each. 1/3 of 51% is 17%. Adding together the 17% that are Bb and the 17% that are bB, we get 34% heterozygous (genotype Bb).

8. **How do manufacturers make vegetable oils solid or semisolid at room temperature?**

A. Adding hydrogen atoms to the double bonds in the fatty acid hydrocarbon chains

B. Removing hydrogen atoms and forming additional single bonds in the fatty acid hydrocarbon chains

C. Removing hydrogen atoms and forming additional double bonds in the fatty acid hydrocarbon chains

D. Adding hydrogen atoms to the single bonds of the fatty acid hydrocarbon chains

Answer: A.

Adding hydrogen atoms to the double bonds in order to form additional single bonds in the fatty acid hydrocarbon chains. The presence of single bonds in the carbon chain will make a fatty acid solid at room temperature because the fatty acids will have fewer "kinks" in their shape, allowing them to stack closer.

9. **If a segment of DNA has a template strand of 5'-TAC GAT TAG-3', what will be the RNA that results from the transcription of this segment?**

A. 3'-TAC GAT TAU-5"

B. 3'-UAC GAU UAG-5'

C. 3'-AUG CUA AUC-5'

D. 3'-ATG CTA ATA-5'

Answer: C.

3'-AUG CUA AUC-5'. When this strand is transcribed, the complementary nitrogen bases will match up, A-U, C-G. Transcription starts at the 5' end of the fragment.

10. **In mammals, nerve cells have spaces between them, across which signals must be transferred. How do these signals get across this gap?**

A. Neural impulses cause the release of chemicals that diffuse across the gap.

B. Sodium and potassium rapidly flux back and forth to carry the signals across the gap.

C. Electrical currents of varying voltages are emitted from one side of the gap to another.

D. The calcium within the axons and dendrites of nerves adjacent to a gap acts as the messenger.

Answer: A.

Neural impulses cause the release of chemicals that diffuse across the gap. When a signal reaches a gap (called a synapse), it stimulates the release of chemicals called neurotransmitters. These chemicals then diffuse across the gap, carrying the signal with them. The neurotransmitters are then received by special receptors on the other side.

11. Why would double fertilization be a useful adaptation in plants?

A. Two sperm fertilizing one egg would provide greater genetic diversity in the offspring, providing a mosaic cellular effect, where different cells of the organisms have different genotypes.

B. Fertilization of a plant by gametes from two separate plants would provide greater genetic diversity in the offspring and reduce the chance for heterozygous dominant alleles causing dysfunctions.

C. Two sperm enter the plant embryo sac; one sperm fertilizes the egg, the other forms the endosperm, which will act as food for the developing embryo.

D. Two sperm enter the animal embryo sac; one sperm fertilizes the egg, the other forms the yolk, which will act as food for the developing embryo.

Answer: C.

Two sperm enter the plant embryo sac; one sperm fertilizes the egg, the other forms the endosperm. In angiosperms, double fertilization is when an ovum is fertilized by two sperm. One sperm produces the new plant and the other forms the food supply for the developing plant (endosperm).

12. The following pedigree illustrates the pattern for an autosomal (non-sex-linked) gene that displays complete dominance. This pedigree only shows affected (shaded) and non-affected individuals.

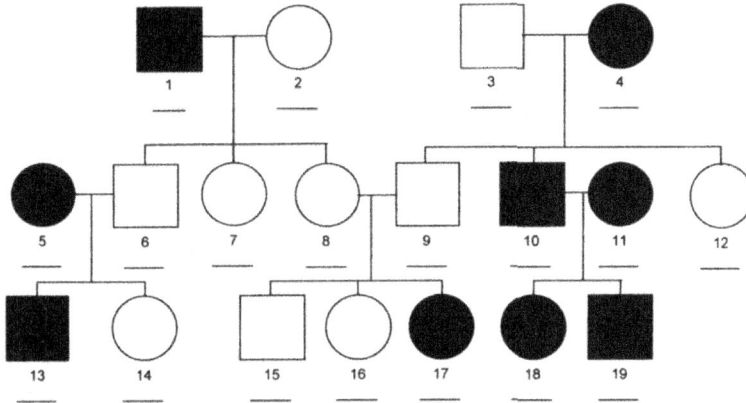

Which members of the above pedigree could be homozygous dominant?

A. 2 and 3

B. 2, 9, and 12

C. 6, 7, and 8

D. 2, 15, and 16

Answer: D.

2, 15, and 16. Person 2 is definitely homozygous dominant. This is clear because all of her children have the dominant phenotype even though their father was homozygous recessive. Persons 6, 7, 8, 9, and 12 are all heterozygous, since one parent was homozygous recessive and they have the dominant phenotype. Person 3 is heterozygous because half of the children have the recessive phenotype, so they must have gotten a recessive gene from their father. Persons 15 and 16 could be either heterozygous or homozygous dominant, since both of their parents were heterozygous.

13. During cellular respiration, there is an enormous transfer of electrons during redox reactions in the mitochondria. What is the result when hydrogen ions are pumped out of the mitochondrial matrix, across the inner mitochondrial membrane, and into the space between the inner and outer membranes?

A. Damage to the mitochondrion

B. The creation of a proton gradient

C. The lowering of pH in the mitochondrial matrix

D. The restoration of the Na-K balance across the membrane

Answer: B.

The creation of a proton gradient. The electron transport chain moves hydrogen ions to one side of the membrane, creating a gradient across the inner and outer membranes. It is the flow of these hydrogen ions across this gradient that ATP synthase uses to convert the energy generated by the flow into bond energy stored in ATP.

14. The relative location of four genes on a chromosome can be mapped from the following data on crossover frequencies:

Gene	Frequency of Crossover
B and D	5%
C and D	50%
C and B	45%
A and B	30%
C and A	15%

Which of the following represents the relative positions of these four genes on the chromosome?

A. ABCD

B. BDCA

C. CABD

D. DBCA

Answer: C.

CABD. The closer two genes are on a chromosome the less likely they are to cross over with each other. Since C and A and B and D have the lowest frequencies, they are next to each other respectively.

15. A scientist investigating the mating behaviors of the zebra finch placed some of its eggs into the nest of another species of finch. They then allowed the zebra finches to be raised by the other species. Results showed that when the zebra finches were ready to start courting females, they turned their attention towards the species of the adoptive parents instead of their own species. Why do the zebra finches behave in this manner?

    A. The zebra finches are acting out of habit.

    B. The zebra finches have imprinted on their new parents.

    C. The other species offers reinforcement techniques to select a mate.

    D. The parents of the other species have conditioned the zebra finches.

    Answer: B.

    The zebra finches have imprinted on their new parents. When the zebra finch eggs hatched, the mother of the other finch species was most likely they first bird the young saw. As a result, they imprinted on her. The zebra finches did not know they were of a different species, so they showed mating behaviors that were common to the species they considered themselves.

16. **What is required when amino acid molecules bond together to form larger molecules?**

    A. The release of water molecules

    B. The increase of activation energy

    C. The addition of water molecules

    D. The release of carbon dioxide molecules

    Answer: A.

    The release of water molecules. The reaction that takes place is called a dehydration synthesis or a condensation reaction. This means that hydrogen and oxygen are released in the form of water in order to bond the rest of the atoms together.

17. **What would a molecule formed from a hydrogen atom (which exhibits weak electronegativity) and an oxygen atom (which has relatively strong electronegativity) demonstrate?**

A. Positive and negative partial charges on the oxygen and hydrogen atoms, respectively.

B. Negative partial charges on both the oxygen and hydrogen atoms.

C. No overall polarity.

D. Negative and positive partial charges on the oxygen and hydrogen atoms, respectively.

Answer: D.

Negative and positive partial charges on the oxygen and hydrogen atoms, respectively. The strong electronegativity of the oxygen atom would draw the electron of the weaker-electronegativity hydrogen atom, creating a dipole in which the oxygen atom exhibits a negative partial charge and the hydrogen atom a positive partial charge. This simple phenomenon forms the basis for the polar nature of water, and hydrogen bonding, an important determinant of macromolecular structures.

18. **The T4 bacteriophage has a double-stranded DNA genome. How might infection with such a virus cause evolutionary change in a bacterial cell?**

A. During replication, a progeny virus incorporates a gene from the infected bacterium into its capsid. It then infects another bacterium and introduces the DNA from the previously infected bacterium into the newly infected bacterium. The newly introduced DNA has no homologous gene on the main chromosome, so it incorporates into the chromosome as an insertion.

B. After infection, the phage genome undergoes reverse transcription. The reverse-transcribed genes of the virus are incorporated into the bacterium's genome.

C. During viral replication, mRNA copies of the infected bacterium's genes are incorporated into progeny virus. When the progeny viruses infect a new bacterium, these genes undergo reverse transcription and are integrated into the newly infected bacterium's genome.

D. During viral replication, a progeny virus incorporates genes from an infected bacterium into its capsid. The progeny virus then introduces the DNA from the previously infected bacterium into a newly infected bacterium and this introduced DNA undergoes homologous recombination with the genome of this newly infected bacterium.

Answer: D.

During viral replication, a progeny virus incorporates genes from an infected bacterium into its capsid. The progeny virus then introduces the DNA from the previously infected bacterium into a newly infected bacterium and this introduced DNA undergoes homologous recombination with the genome of this newly infected bacterium. The two ways that bacterial genes can be transferred to other bacteria is by homologous recombination (D) or if the DNA was a plasmid. If it does not have a homologous gene in the recipient bacterium, it will simply be recycled in the second bacterium as a nutrient, so A is incorrect; it will not affect evolution. B and C are incorrect because a double-stranded DNA genome virus does not use reverse transcriptase.

19. **The fossil record shows that large flying insects arose approximately 250 million years ago. Evidence also suggests that the oxygen concentration of the atmosphere at this time was approximately 28% higher compared to present day levels. Knowing this, what physiological features allows mammals to attain great size, whereas today's insects remain relatively small?**

    A. Red blood cells and the mammalian immune system.

    B. White blood cells and the mammalian immune system.

    C. Lungs and the mammalian circulatory system.

    D. The heart and the mammalian central nervous system.

    Answer: C.

    Lungs and the mammalian circulatory system. These are complex and allow oxygen to be delivered to tissues far away from the air. Insects, without this physiology, are restricted in size because they would not be able to deliver sufficient oxygen at larger sizes.

20. **A retrovirus infects a cell. Retroviruses have RNA genomes. How can this infection cause evolutionary change in the species?**

    A. mRNA is transcribed from the retroviral genome and this mRNA is reverse transcribed and then incorporated into the host genome using integrase.

    B. Genes from the retroviral genome are directly incorporated into the host cell genome using homologous recombination.

    C. Host mRNA and genes from the retroviral genome are reverse transcribed and then incorporated into the host genome using integrase.

    D. Retroviral proteins are converted into DNA by protease, and then integrated into the host genome using integrase.

    Answer: C.

    Host mRNA and genes from the retroviral genome are reverse transcribed and then incorporated into the host genome using integrase. Reverse transcriptase converts RNA into DNA and integrase integrates the DNA into the host's DNA genome. A is incorrect because mRNA cannot be made from the retroviral genome since the retroviral genome is RNA. B is also incorrect because the retroviral genome is RNA; they cannot be incorporated into a DNA genome. D is incorrect because protease does not convert proteins into DNA. (It cuts large polyprotein polypeptides into individual proteins or peptides.)

21. **cDNA is an mRNA sequence that has been reverse transcribed into DNA. Which of the following could be true?**

   A. cDNA can be created by a virus; it will be an exact copy of the gene from which the mRNA was transcribed, causing a point mutation.

   B. cDNA can be created by the use of reverse transcriptase by molecular biologists; it will be larger than the gene from which the mRNA was transcribed.

   C. cDNA can be created by either a virus or a molecular biologist using reverse transcriptase; it will be an exact copy of the gene from which the mRNA was transcribed.

   D. cDNA can be created by either a virus or a molecular biologist using reverse transcriptase; it will be smaller than the gene from which the mRNA was transcribed.

Answer: D.

cDNA can be created by either a virus or a molecular biologist using reverse transcriptase; it will be smaller than the gene from which the mRNA was transcribed. When mRNA is created, the introns in a gene are spliced out. (It also does not include regulatory sequences such as the promoter.) When the exon-only mRNA is reverse transcribed, it will be substantially smaller than the intron-containing original gene. (You should be able to eliminate A quickly, because the mutation would not be a point mutation.)

22. **The worldwide population of the North Atlantic right whale has dramatically decreased, almost to the point of extinction. It also shows very little genetic diversity. Which of these is the most important risk factor the whales face as a result of this?**

   A. The habitat of the whales is threatened by climate change and global warming.

   B. A dwindling food supply makes it more difficult for the right whales to survive and reproduce.

   C. Mutations are more likely to affect more individuals in populations with low genetic diversity.

   D. The low genetic diversity interferes with the ability of the population to respond to environmental changes.

Answer: D.

The low genetic diversity interferes with the ability of the population to respond to environmental changes. Since the population of whales is so low, it shows a very low genetic diversity. As a result, the gene pool for new traits is quite small. If a change in the environment occurred that required an adaptation to survive, it is unlikely the whales would be able to accommodate it.

23. **An antiviral drug inhibits the protein reverse transcriptase. Against what type of virus will the drug be effective?**

    A. Bacteriophage T4, a lytic phage with a double-stranded DNA genome

    B. HIV, a lentivirus with a double-stranded DNA genome

    C. Parvovirus B19, a parvovirus with a single-stranded DNA genome

    D. SIV, a lentivirus with a single-stranded RNA genome

    Answer: D.

    SIV, a lentivirus with a single-stranded RNA genome. Reverse transcriptase is only used by retroviruses, with RNA genomes. (While B may be tempting because HIV is a well-known retrovirus inhibited by reverse transcriptase inhibitors, HIV does not have a double-stranded DNA genome.)

24. **Algae, including seaweed, and plants are similar in that they are both eukaryotic organisms that create their energy from sunlight through photosynthesis. Which of the following are fundamental differences between algae and plants?**

    A. The evolution of xylem in plants has allowed them to become much larger than algae.

    B. Flowers allow plants to be fertilized by insects; this kind of symbiotic relationship is not seen with algae.

    C. In the ocean, swimming gametes can meet, as sperm cells meet egg cells. On land, seeds are needed to allow fertilization to occur.

    D. Land plants require tissue differentiation that is not necessary in algae.

    Answer: D.

    Land plants require tissue differentiation that is not necessary in algae. While some algae have specialized anatomical structures, this differentiation is not fundamentally required. On land, however, plants must differentiate since part of the plant must be underground in order to access water and minerals, and part of the plant must be above ground to access sunlight. A is incorrect; algae can be very large—some kelp can grow to 300 feet in height, and plants can be very small and nonvascular. B is incorrect; not all plants are flowering plants, and many algae form symbioses with other organisms. C is incorrect because, while most plants have evolved seeds or other structures to allow fertilization in dry conditions, plants like mosses still have swimming sperm and egg gametes. They are, of course, dependent on moist conditions for reproduction. Simple plants like mosses are not nearly as differentiated as higher plants (with fully separated functions in roots, stems, and leaves) but they still require tissue differentiation that is not necessary in algae.

25. **A scientist wants to find fossils from dinosaurs that lived at the end of the Jurassic period. Where would be a reasonable place to look?**

A. Igneous rocks from the late Triassic period (the era just preceding the Jurassic)

B. Sedimentary rocks from the late Triassic period (the era just preceding the Jurassic)

C. Igneous rocks from the early Cretaceous period (the era just after the Jurassic)

D. Sedimentary rocks from the early Cretaceous period (the era just after the Jurassic)

Answer: D.

Sedimentary rocks from the early Cretaceous period (the era just after the Jurassic). Fossils would be expected in sedimentary rocks, which form from layers of sand/silt being deposited over time. Igneous rocks would rarely be expected to contain fossils since they form from molten rock. Sedimentary rocks from the late Triassic would already have formed by the time Jurassic organisms lived. Sedimentary rocks from the Cretaceous, however, would be expected to have formed around animals that died at the end of the Jurassic.

26. **Myoglobin is located in the muscles. Which of the following are true about myoglobin?**

A. Myoglobin has a higher affinity for oxygen than hemoglobin because muscles need to receive oxygen from blood.

B. Hemoglobin has a higher affinity for oxygen than myoglobin because of the importance of transporting oxygen throughout the body.

C. Hemoglobin has a higher affinity for oxygen than myoglobin because it needs to also carry CO_2, a waste product of respiration.

D. Myoglobin has a higher affinity for oxygen than hemoglobin because muscles sometimes need to undergo anaerobic respiration.

Answer: A.

Myoglobin has a higher affinity for oxygen than hemoglobin because muscles need to receive oxygen from blood. If myoglobin did not have a higher affinity for oxygen than hemoglobin, the blood would never be able to pass the oxygen on for the muscles to use!

27. **Which of the following could be expected to lead to anemia in a mammal, and why?**

    A. A deficiency of iron, because iron is important for functional heme

    B. A deficiency of iron, because iron carries oxygen into the nucleus of red blood cells

    C. Genetic disorders that cause abnormal hemoglobin, because this will result in a lower number of red blood cells

    D. Genetic disorder that cause a higher-than-normal number of red blood cells, but without any functional hemoglobin

    Answer: A.

    A deficiency of iron, because iron is important for functional heme. Iron is the center of heme, and is responsible for heme's ability to carry oxygen. B is wrong because oxygen is not carried in the nuclei even in animals that have nucleated red blood cells; furthermore, in mammals, red blood cells have no nuclei. C is wrong because abnormal hemoglobin will not cause the *number* of red blood cells to decrease; it will cause their oxygen-carrying efficiency to decrease. D is wrong because if there was no functional hemoglobin, the mammal would not be alive.

28. **In mosses, the dominant part of the plant is the gametophyte, the velvety green part that looks a bit like a miniature lawn. Gametes are formed from this part by mitosis. What is true about this velvety, dominant stage of the moss?**

    A. It is haploid; this differentiates mosses from vascular plants, in which the sporophyte is dominant.

    B. It is diploid; this differentiates mosses from vascular plants, in which the sporophyte is dominant.

    C. It is haploid, as is the sporophyte in vascular plants

    D. It is diploid, as is the gametophyte in vascular plants

    Answer: A.

    It is haploid; this differentiates mosses from vascular plants, in which the sporophyte is dominant. The dominant stage in mosses is the haploid gametophyte. Gametes are always haploid; in order to make haploid gametes by mitosis, you must have a haploid parent. (You could either know this about mosses, or you could deduce this from the statement about mitosis in the question.)

29. **The lack of a nucleus in many vertebrate red blood cells has been proposed as a reason for the accumulation of "junk" or vestigial non-coding DNA in those genomes. What evolutionary factors support this view?**

A. The nucleus helps to control the amount of non-coding DNA; without nuclei in their red blood cells, these vertebrates are unable to "clean up" junk DNA.

B. Not having nuclei in red blood cells saves energy that can be used in other cells to maintain junk DNA.

C. In organisms with nuclei in their red blood cells, the cells are bigger, so in order to fit in small blood vessels, there is selection pressure to keep the nuclei smaller. In animals that do not have nuclei in their red blood cells, larger nuclei do not affect the size of the red blood cells, so this selection pressure is absent.

D. In organisms with nuclei in their red blood cells, the efficiency of oxygen transport is greater, so there is greater selection pressure to retain their nuclei. This selection pressure is absent in animals that do not have nuclei in their red blood cells, allowing the accumulation of junk DNA.

Answer: C.

In organisms with nuclei in their red blood cells, the cells are bigger, so in order to fit in small blood vessels, there is selection pressure to keep the nuclei smaller. In animals that do not have nuclei in their red blood cells, larger nuclei do not affect the size of the red blood cells, so this selection pressure is absent. A is wrong because, even if this were true, red blood cells are not reproductive cells, so even if those nuclei had some cleanup capability, this would not contribute to cleaning up the DNA in the germline. B is wrong because this is not the way energy works in the body. D is wrong—among other reasons—because nuclei have nothing to do with efficiency of oxygen transport; hemoglobin is located in the cytoplasm.

30. **Research has demonstrated that some cancer patients with a poor prognosis have high levels of a specific receptor tyrosine kinase called HER2. Which of the following reasons would explain why creating a drug to bind HER2 might inhibit tumor growth?**

A. Binding HER2 would increase the phosphorylation of downstream proteins.

B. Binding HER2 would prevent the phosphorylation of downstream proteins.

C. Binding HER2 would allow the flow of ions through the membrane.

D. Binding HER2 would block the flow of ions through the membrane.

Answer: B.

Binding HER2 would prevent the phosphorylation of downstream proteins. Receptor tyrosine kinases are not ion-gated channels therefore C and D are incorrect. Increasing the phosphorylation of downstream targets would facilitate signaling and growth of the tumor, therefore, inhibiting the receptor and downstream signaling could potentially stop tumor growth.

31. **A flower is most commonly pink, but occasionally white or yellow versions are seen. After several crosses, the following data were obtained:**

Parents	F1	F2
Pink x white	Pink	195 Pink, 64 white
Pink x yellow	Pink	150 Pink, 53 yellow
white x yellow	Pink	228 Pink, 97 white, 76 yellow

Which of the following statements best explains the data?

A. The presence of pink in all three groups of the F1 generation indicates that this phenotype is a function of the environmental conditions present during the F1 cross that are different from those during the F2 cross

B. Pink and yellow are inherited, segregated genes that follow Mendel's laws, but white is a function of the flowers not having matured enough for their color to show

C. Pink and yellow follow Mendel's laws but white demonstrates non-Mendelian inheritance; white probably results from vegetative (clonal) reproduction

D. Pink, white, and yellow are all inherited genes. The inconsistency could be due to a small sample size, or it could be due to some other factor, such as another gene influencing the phenotypic expression of the flowers.

Answer: D.

Pink, white, and yellow are all inherited genes. The inconsistency could be due to a small sample size, or it could be due to some other factor, such as another gene influencing the phenotypic expression of the flowers. The appearance of a single, dominant trait in the F1 generation is typical of Mendelian inheritance. The subsequent ratios are approximately those expected in Mendelian inheritance (75:25 for complete dominance; 25:50:25 for codominance), but experimental data will never give those exact ratios, just as a series of coin tosses will never give a 50:50 ratio. The larger the sample size, the closer the ratio will be to the theoretical one, but there will always be an element of random variation. In addition, another factor, such as another gene influencing the color produced, could affect the phenotype.

32. After doing several more crosses (for the same flowers from the previous question), the data look like this:

Parents	F1	F2
Pink x white	Pink	740 Pink, 230 white
Pink x yellow	Pink	760 Pink, 270 yellow
white x yellow	Pink	480 Pink, 260 white, 240 yellow

Which of the following statements can now be made with certainty?

A. The traits are inherited, pink is dominant, and the data conform to Mendel's law of independent assortment.

B. The traits are inherited, pink and yellow are codominant, and the data conform to Mendel's law of independent assortment.

C. The traits are not inherited.

D. The traits are inherited, pink is dominant, and the data conform to Mendel's law of segregation.

Answer: D.

The traits are inherited, pink is dominant, and the data conform to Mendel's law of segregation. With a larger sample population, the new F2 numbers conform very closely to a 75%/25% ratio for dominant/recessive traits and a 50%/25%/25% ratio for codominant traits. The tricky thing is that the *genotype* of the F1 pink phenotype for the white x yellow cross is different from the genotype of the F1 pink phenotype for the Pink x white and Pink x yellow crosses. Because it is due to the interaction of the white and yellow alleles, rather than presence of the Pink allele, it will probably be a different shade of pink.

33. **A patient is diagnosed with a rare viral infection that has produced symptoms that include a severe cough that causes hemorrhaging of the lungs. The coughed up blood contains viable virus particles, but conventional PCR for DNA does not produce any visible bands on an electrophoresis gel. Why are the viral particles not likely to be detected with this method?**

A. A protein coat surrounding the nucleic acid prevents the DNA from amplifying.

B. The virus is an RNA virus and would need reverse transcription to be detected by amplification.

C. The surrounding a protein coat only contains the base plates and the tails of the virus.

D. The DNA concentration is too low in the blood samples for PCR amplification.

Answer: B.

The virus is an RNA virus and would need reverse transcription to be detected by amplification. RNA viral particles are packaged RNA molecules inside the protein coat.

34. The graph shows the results of an investigation into the amount of energy (in the form of heat) produced by different sized animals.

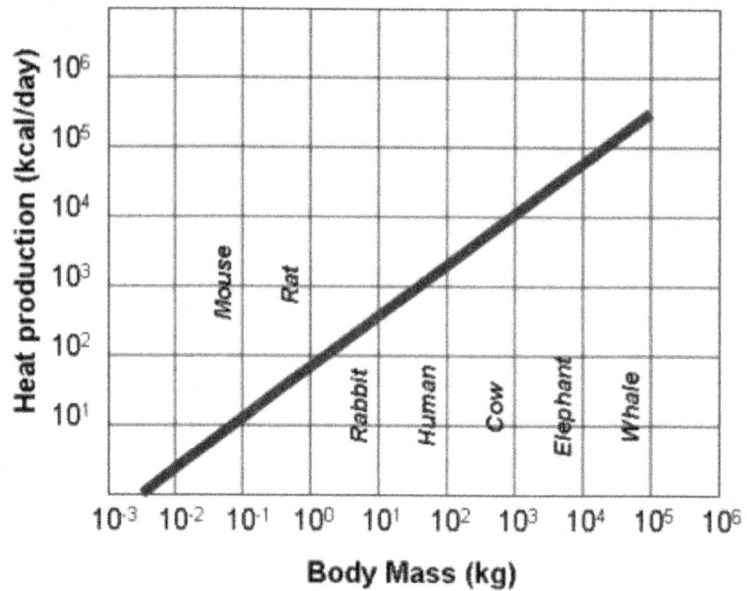

What conclusion can be made from the information provided in the graph?

A. Animals that have a large body mass also produce large amounts of heat.

B. Body mass is inversely related to the amount of heat produced by an animal.

C. The individual cells of small animals produce less heat than those of large animals.

D. Larger animals possess adaptations for dissipating heat more effectively than smaller animals.

Answer: A.

Animals that have a large body mass also produce large amounts of heat. While answer D is true, this cannot be concluded from the data in the graph. The graph only deals with the correlation between body mass and heat production, not heat-dissipating adaptations.

35. **Hyperventilation can cause the pH of the blood to rise. Considering this chemical equation:**

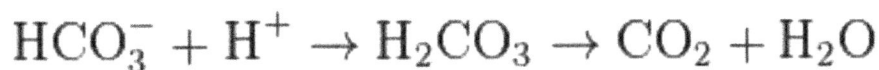

$$HCO_3^- + H^+ \rightarrow H_2CO_3 \rightarrow CO_2 + H_2O$$

hyperventilation will result in which of the following?

A. Respiratory acidosis, because CO_2 is released in a greater amount, driving this reaction to the right and making the blood more acidic.

B. Respiratory alkalosis, because CO_2 is released in a greater amount, driving this reaction to the right and making the blood more basic.

C. Respiratory acidosis, because CO2 is released in a greater amount, driving this reaction to the left and increasing the concentration of hydrogen ions.

D. Respiratory alkalosis, because CO_2 is released in a greater amount, driving this reaction to the left and increasing the amount of H_2CO_3.

Answer: B.

Respiratory alkalosis, because CO_2 is released in a greater amount, driving this reaction to the right and making the blood more basic. A higher pH means the blood is becoming more alkaline (alkalosis). The release of CO_2 drives the reaction to the right, which will cause a decrease in the acid H_2CO_3 and therefore a decrease in the concentration of H^+ ions, which means a higher pH.

36. The graph shows the rates of survivorship for different organisms.

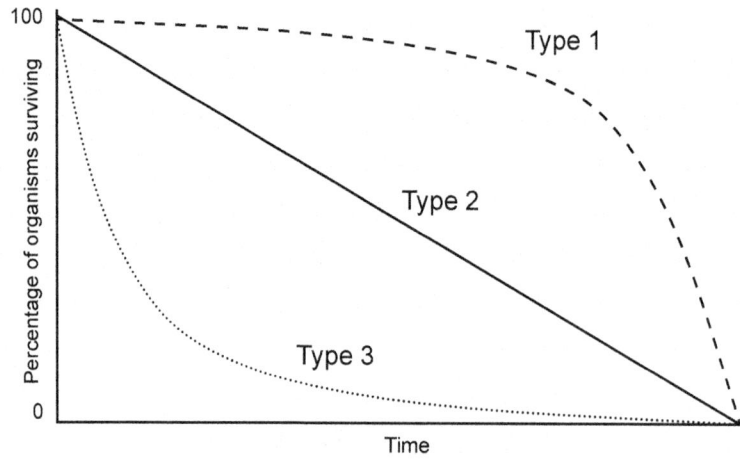

Which of the following pairs best matches the expected curves for field mice and humpback whale?

A. Field mouse – Type III; humpback whale – Type I

B. Field mouse – Type I; humpback whale – Type III

C. Field mouse – Type II; humpback whale – Type II

D. Field mouse – Type I; humpback whale – Type I

Answer: A.

Field mouse – Type III: humpback whale – Type I. Those organisms with Type III survivorship produce a lot of offspring at first, but they do not live for a long time. These are usually prey species, such as mice and other rodents. Insects also fall into this category. Animals with Type I survivorship, like the humpback whale, produce few offspring, but since the investment into them is so great, they tend to live for a long time.

37. Using the following taxonomic key, identify the tree from which this branch came.

1 - Are the leaves PALMATELY COMPOUND (BLADES arranged like fingers on a hand)? – go to 2

1 - Are the leaves PINNATELY COMPOUND (BLADES arranged like the veins of a feather)? – go to 3

2 - Are there usually 7 BLADES - *Aesculus hippocastanum*
2 - Are there usually 5 BLADES - *Aesculus glabra*

3 - Are there mostly 3-5 BLADES that are LOBED or coarsely toothed? - *Acer negundo*
3 - Are there mostly 6-13 BLADES with smooth or toothed edges? - *Fraxinus americana*

    A. *Aesculus hippocastanum*

    B. *Aesculus glabra*

    C. *Acer negundo*

    D. *Fraxinus americana*

Answer: C.

Acer negundo. The leaves are pinnately compound, with 5 coarsely toothed leaves, leading to the answer: *Acer negundo.* The list below includes the scientific name and the common name for the all the plants listed above:

Aesculus hippocastanum (Horsechestnut)
Aesculus glabra (Ohio Buckeye)
Acer negundo (Boxelder, Ashleaf Maple)
Fraxinus americana (White Ash)

38. **It is thought that mitochondria evolved from symbiotic bacteria that were engulfed by a larger cell. Which of the following represents evidence that supports this?**

    A. Mitochondria are smaller than the eukaryotic cells; eukaryotic cells have linear chromosomes.

    B. Mitochondria have their own DNA; eukaryotic genomes are phylogenetically similar to the archaea.

    C. Mitochondria encode ATP synthase; eukaryotic cells encode the machinery for glycolysis.

    D. Mitochondria have double membranes; eukaryotic genomes are phylogenetically similar to the eubacteria.

Answer: B.

Mitochondria have their own DNA; eukaryotic genomes are phylogenetically similar to the archaea. The presence of a separate DNA genome is strong evidence that mitochondria evolved independently, as this is highly unlikely to have occurred if the eukaryotic cell had simply evolved the organelle as a structure. The phylogenetic similarity of eukaryotic and archaean genomes is strong evidence that the main cell is of non-eubacterial origin, so this indicates that a bacterium did not simply expand, developing a large outer coat that eventually evolved into the main cell. D is tempting, because the double membrane practically forms a picture of a small cell engulfed by a large one, but it is not true that eukaryotic genomes are phylogenetically similar to eubacteria, and also, if it were true, it would contradict the endosymbiosis theory, not support it.

39. **Chloroplasts are thought to have evolved from endosymbiosis of photosynthetic cyanobacteria. Which of the following represents evidence that supports this?**

    A. Chloroplasts conduct photosynthesis using H_2O and light; chloroplasts are found only in plants

    B. Cyanobacteria conduct photosynthesis using H_2O and light; some free-living chloroplasts persist in remote environments

    C. Chloroplasts and cyanobacteria both have thylakoid membranes; the genome sequences of chloroplasts are homologous to those of cyanobacteria

    D. Cyanobacteria conduct photosynthesis using O_2 and light; chloroplasts have thylakoid membranes like cyanobacteria

Answer: C.

Chloroplasts and cyanobacteria both have thylakoid membranes; the genome sequences of chloroplasts are homologous to those of cyanobacteria. A is incorrect because chloroplasts are not only found in plants; they are also found in algae and cyanobacteria. B is incorrect because there are no free-living chloroplasts, although cyanobacteria do resemble chloroplasts. D is incorrect because photosynthesis is not conducted using O_2 and light; O_2 is a product of photosynthesis, not a reactant. C is correct because both of these points support the theory and both of these points are true.

40. Yellow squash and zucchini plants sometimes spontaneously mutate to become toxic, even deadly. Plants evolve in this way to protect themselves from being eaten. What difference between plants and animals underlies this evolutionary pattern?

A. Plants lack immune systems.

B. Plants lack digestive systems.

C. Plants lack nervous and musculoskeletal systems.

D. Plants can mutate easier because they can reproduce clonally.

Answer: C.

Plants lack nervous and musculoskeletal systems. Because plants lack the behavioral ability afforded by the nervous system and musculoskeletal system of animals, they cannot run from or fight herbivores. They have therefore evolved other methods of defense, including the production of toxins, as well as bad-tasting chemicals and chemicals that cause pain, such as in hot peppers.

41. A woman has Pearson Syndrome, a disease caused by a mutation in mitochondrial DNA. In which of the following individuals would you expect to see the disease?

I   Her daughter
II Her son
III Her daughter's son
IV Her son's daughter

A. I, III

B. I, II, III

C. II, IV

D. I, II, III, IV

Answer: B.

I, II, III. Since mitochondrial DNA is passed through the maternal line, both of her children would be affected and the trait would continue to pass from her daughter to all of her children. Her son's children would recieve their mitochondrial DNA from their mother.

42. **Photosynthetic organisms like plants and algae use H_2O as a reactant. However, some deep-sea organisms survive in the complete absence of light. Which two things could be true about such organisms?**

A. They could produce their own light through bioluminescence or they could be heterotrophs.

B. They could be heterotrophs or they could be algae that translocate between light and dark levels of the ocean.

C. They could be oligotrophs that use sulfur from deep ocean vents as energy, or they could be heterotrophs.

D. They could be heterotrophs or they could be autotrophs that use H_2S as a reactant.

Answer: D.

They could be heterotrophs or they could be autotrophs that use H_2S as a reactant. A is incorrect because production of light through bioluminescence is a means of utilizing energy, not obtaining energy. B is incorrect because algae do not do this; the distance would be far too great and the difference in pressure would preclude the same organisms from living in both levels. C is incorrect because while oligotrophs need only very low levels of nutrients, they still need nutrients—the term refers to the amount of nutrients they need (they can live in nutrient-poor environments), not to their being able to produce energy as autotrophs can. Further, sulfur cannot be used as energy; autotrophs can harness the energy released by the breakdown of H_2S, to make high-energy compounds like glucose, just as photosynthetic autotrophs do.

43. **Which of the following would be considered an example of apoptosis that occurs during the embryological development of an animal?**

A. An increase in the growth rate of bone forming cells that increase bone length in the animal.

B. The death and disintegration of certain cells in the developing limb to form fingers and toes.

C. A decrease in the rate of cell division of liver tissue to ensure development does not occur too quickly.

D. The activation of regulatory genes in the cells that control the contraction and relaxation of the developing heart.

Answer: B.

The death and disintegration of certain cells in the developing limb to form fingers and toes. Apoptosis is programmed cell death. In the developing embryo, cells of the appendages undergo this process to remove certain cells. When this happens, the digits separate.

44. **The archaea are prokaryotic organisms with genomes that are very different from the genomes of eubacteria, and more similar to eukaryotic genomes. They tend to be found in extreme environments, such as extremely salty (halophiles), extremely hot (thermophiles), and extremely low pH (acidophiles). What might explain this phenomenon?**

A. Archaea were originally ubiquitous, but when some archaean ancestors formed symbioses with eubacteria to become eukaryotes, they outcompeted the archaea. Only those archaea who could handle extreme conditions survived, and the traits that allowed this were selected by natural selection.

B. Archaea have always been extremophiles, since life was created. The original world contained archaea, eubacteria, and eukaryotes, but the eukaryotes expanded because of their superior survival capabilities; this is an example of natural selection after creation.

C. All original life forms were eubacteria, and archaea evolved from them in extreme environments. Evidence for this is that they are both prokaryotes; eukaryotes have a nucleus and organelles, cell structures that evolved independently from eubacteria in mild conditions at a later date.

D. Eubacteria were originally ubiquitous, but then some formed endosymbiotic relationships, engulfing archaea to become eukaryotic organisms capable of respiration. The archaea that lived in extreme environments, however, were not in contact with eubacteria and escaped the fate of becoming organelles.

Answer: A.

Archaea were originally ubiquitous, but when some archaean ancestors formed symbioses with eubacteria to become eukaryotes, they outcompeted the archaea. Only those archaea who could handle extreme conditions survived, and the traits that allowed this were selected by natural selection. Early in the evolution of life, there were archaea and eubacteria, both prokaryotes. When some archaeans engulfed eubacteria and formed an endosymbiosis, leading to the eukaryotic cell, they gained a huge competitive advantage over other archaeans. The only archaeans that survive today are those that were able to adapt to environments were eukaryotic cells cannot survive, such as highly salty lakes and geysers where the water reaches over 100°C. B is not correct because life was not created and the original world did not contain eukaryotes. C is incorrect because it is not true and does not explain the phenomenon. D is incorrect because eubacteria did not engulf any cells to become eukaryotes; eukaryotic cells engulfed bacteria. This idea of a whole population of cells eating another population, and some escaping, is also not in line with evolution. The pattern is instead that one cell engulfed another, and then reproduced to form a progeny population.

45. **Mitochondrial genomes are much smaller than nuclear genomes, and most of the proteins used by mitochondria are encoded in the nucleus and subsequently transported to the mitochondria. Many of these nuclear genes are phylogenetically similar to genes from bacteria. Why might this be?**

A. The nucleus of the original eukaryotic cell encoded proteins needed by the mitochondrion; this is why the evolutionary event of endosymbiosis occurred.

B. At the time of the ancient endosymbiosis event, the mitochondria encoded all of their genes, but some genes were transferred from the mitochondria to the nucleus through lateral gene transfer.

C. At the time of the ancient endosymbiosis event, the mitochondria encoded all of their genes, but then, the mitochondria began using nuclear genes, and the mitochondrial genes were eliminated from the mitochondria to avoid redundancy.

D. The nuclear genome evolved superior genes that accomplished the functions of the original mitochondrial genes, but with greater efficiency. This is an example of convergent evolution.

Answer: B.

At the time of the ancient endosymbiosis event, the mitochondria encoded all of their genes, but some genes were transferred from the mitochondria to the nucleus through lateral gene transfer. Explanation: In A, C, and D, the genes would not be phylogenetically homologous with bacterial genes; B is the only scenario of the four that would give rise to bacterial genes in the nuclear genome. This is an important piece of evidence for the existence of lateral gene transfer.

46. **The diagram represents two nerve cells that are adjacent to each other. Which sequence of events correctly describes the transmission of the nerve impulse as it travels from point A to point E?**

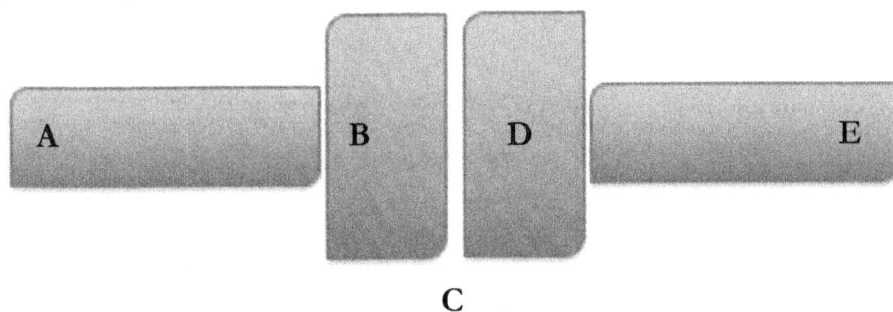

A. At point A, an influx of Na and outflow of K initiates the action potential, which travels to the end of the axon (B). Neurotransmitters are released into the synapse (C) and are absorbed by the next axon (D). The neurotransmitter initiates an action potential, which travels to point E.

B. At point A, an influx of Na and outflow of K initiates the action potential, which travels to point B. The synapse (C) is flooded with Na that is absorbed by the next axon (D), thus continuing the action potential to point E.

C. At point A, an influx of neurotransmitters initiates the action potential, which travels to point B. Na is released into the synapse (C) and is absorbed by the next axon (D). The Na stimulates the influx of neurotransmitters in the next axon and the action potential travels to point E.

D. At point A, an influx of Na stimulates the release of neurotransmitter within the axon cell. The neurotransmitters flow to the end of the axon (B), cross the synapse (C), and continue to point E.

Answer: A.

At point A, an influx of Na and outflow of K initiates the action potential, which travels to the end of the axon (B). Neurotransmitters are released into the synapse (C) and are absorbed by the next axon (D). The neurotransmitter initiates an action potential, which travels to point E. The signal that transmits across a neuron is due to the electrical impulse called an action potential. The signal starts with the influx of sodium and the outflow of potassium. As this happens across the nerve membrane, the signal gets carried. When it reaches the end of the axon, it stimulates the release of neurotransmitters, which carry the signal across the synapse. Receptors on the dendrites of the adjoining neuron pick up the signal and continue its transmission through the next cell.

47. **A plant was grown in a test tube containing radioactive nucleotides, the molecules from which DNA is built. When examining the dividing cells in the plant after several days of growth, where would the researchers expect the majority of the radioactivity to be concentrated?**

    A. Rough endoplasmic reticulum

    B. Smooth endoplasmic reticulum

    C. Central vacuole

    D. Nucleus

    Answer: D.

    Nucleus. The replication of DNA in cell division occurs in the nucleus, therefore, the radioactive nucleotides would be incorporated to the DNA, which resides in the nucleus.

48. **A man and a woman both claim to be distant descendants of a Saudi Arabian king who lived over a century ago. A DNA test reveals that the man is definitely his descendent, but no such statement can be made for the woman. Why can paternity be more definitively proven with the son than with the daughter?**

    A. Since women have two X chromosomes, the DNA from one interferes with the DNA from the other, preventing sequencing.

    B. Only men have a Y chromosome, and this is a smaller chromosome that encodes fewer genes, allowing less ambiguity when comparing the sequences.

    C. Only men have a Y chromosome, precluding recombination, so a descendant through the male line will have an almost exact copy of the Y chromosome sequence of his ancestor.

    D. While the sequence of one X chromosome of a woman comes from her father, the genes are inactivated though X-inactivation and this epigenetic modification prevents a definitive comparison.

    Answer: C.

    Only men have a Y chromosome, precluding recombination, so a descendant through the male line will have an almost exact copy of the Y chromosome sequence of his ancestor. As with autosomes, the two X chromosomes undergo recombination during crossing over in women. However, the Y chromosome cannot undergo recombination because it is so unique; it therefore retains its identity and except for any potential mutations, it will be an exact copy of the ancestral patrilineal forefather's Y chromosome, even after many generations.

49. **Some studies have suggested that smallpox is more severe in people with type A blood than people with type O or type B blood. Why might this be?**

A. Type A blood cells have membranes that are more vulnerable to viral entry than those of other blood types

B. The smallpox virus encodes anti-type A antibodies that attack the blood of type A humans.

C. The antigen on the nuclear membrane of type A blood cells binds the smallpox virus, facilitating viral entry.

D. The smallpox virus has surface antigens that resemble type A blood, allowing it to "hide" in humans with type A blood.

Answer: D.

The smallpox virus has surface antigens that resemble type A blood, allowing it to "hide" in humans with type A blood. Explanation: This would be the most likely explanation of the four: A is incorrect because only the antigenic tags of different blood types are different; not the membranes. B is incorrect because viruses cannot encode antibodies; antibodies are produced in a highly complex process of honing the immune system during development of a higher animal. C is incorrect because antigens are not found on the nuclear membrane.

Items 50 - 54 refer to the following graph:

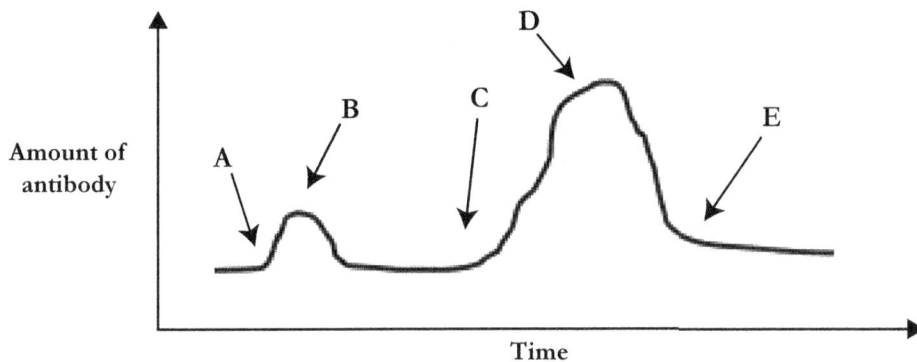

50. **Which of the points above might represent a vaccination and a booster shot?**

A. A and C

B. B and D

C. A and B

D. C and D

Answer: A.

A and C. After giving a vaccination, antibodies rise and then fall, but the immune system is strengthened for the next exposure to an antigen. A booster shot will therefore cause the body to make even more antibodies.

51. What might point C in the graph represent?

A. Exposure to a viral infection after vaccination against the virus

B. Exposure to a viral infection after a previous exposure to the virus

C. Both A and B

D. Neither A nor B; antibodies are only produced in response to bacteria

Answer C.

Both A and B. The event at point A was either an infection or a vaccination. It led to the creation of antibodies and memory cells. At point C, something happens that causes a larger burst of antibody creation at point D. That something would be the presentation of an antigen—either through natural infection or through a second vaccination. In order to get to point D, the event at point A needed to cause the response at point B. That event could either be a vaccination or previous natural exposure to the virus. A and B are both possible, but C is the best answer. D is incorrect because it is not true.

52. If the graph represents response to a viral infection, why is point D higher than point B?

A. The antibodies produced at point B have been supplemented by additional antibodies at point D; point D represents the sum of the antibodies produced in response to the antigen at point A and the antigen at point C.

B. The infecting virus at point C was stronger; for example, cowpox and smallpox both produce the same kinds of antibodies, but smallpox induces more antibodies than cowpox.

C. The person ingested zinc and vitamin C, known immune boosters, at point C, which induced the production of antibodies.

D. The antibodies at point B were produced and then levels returned to normal; however, the immune system was primed by the exposure at point A to be capable of efficiently producing a greater number of antibodies in response to the antigen at point C.

Answer: D.

The antibodies at point B were produced and then levels returned to normal; however, the immune system was primed by the exposure at point A to be capable of efficiently producing a greater number of antibodies in response to the antigen at point C. This is how the immune system works. An exposure to an antigen, such as a virus, causes the production of antibodies. When the infection is over, the antibodies decline, but cells called memory cells stay in place. These cells are capable of quickly generating a large response against any subsequent infection with an antigen. This is why vaccination using a weak version of a virus or similar less virulent virus works—the subsequent response will be strong enough to fight off the real, disease-causing full-strength virus because it is a stronger response than that produced against the weak (attenuated) virus.

53. **What might the graph look like if the virus were to once again infect this person shortly after point E?**

A. A peak similar in size to point D would be observed. The same B cells that recognized the antigen at point C would recognize a subsequent exposure to the antigen.

B. A peak twice as large as that at point D would be observed; every exposure will double the immune response since more antibodies are being added each time.

C. A smaller peak than that seen at point D would be observed, because the immune system would have become habituated to the antigen.

D. No peak would be observed, because the antibodies would be used up and would need time to build up again.

Answer: A.

A peak similar in size to point D would be observed. The same B cells that recognized the antigen at point C would recognize a subsequent exposure to the antigen. The memory of an infection/antigen exposure is at the level of the B cell that produces the antibodies, not the antibodies themselves, so there is no issue of antibodies either adding up or being used up.

54. **Babies receive antibodies through their mother's milk. This is known as passive immunity. If a baby is exposed to an antigen and the mother's antibodies fight off the infection, what might a subsequent infection look like after the baby has been weaned?**

A. It will look the same as a vaccination; whether the baby fought off the infection through the mother's antibodies or its own antibodies after vaccination makes no difference in the response.

B. It will be a much smaller response than if the baby had been vaccinated. This is because a vaccination induces the production of memory cells in the baby whereas passive immunity does not.

C. It will be much larger than if the baby had been vaccinated because the passive immunity produces a population of memory cells and the memory cell population will increase with exposure to the infection.

D. There will be no response. Passive immunity causes a baby's immune system to become dependent on the mother's antibodies, so that after weaning, it is too weak to produce its own.

Answer: B.

It will be a much smaller response than if the baby had been vaccinated. This is because a vaccination induces the production of memory cells in the baby whereas passive immunity does not. The baby was protected by its mother's antibodies, so it did not develop a full infection that would have led to the normal development of memory cells. The disease was fought off, but through the mother's efforts, not the baby's. There may be a somewhat enhanced response to subsequent infection, since the baby was still exposed to some antigen, but not nearly the magnitude of the response that would have been seen if the baby had been vaccinated.

55. In humans, the hormone testosterone enters cells and binds to specific proteins, which in turn bind to specific sites on the DNA of the cell. What is the function of these proteins?

A. Help RNA polymerase transcribe certain genes

B. Promote recombination

C. Unwind the DNA so that its genes can be transcribed

D. Cause mutations in the DNA

Answer: A.

Help RNA polymerase transcribe certain genes. Testosterone binds to androgen receptor, which is a transcription factor. When androgen receptor binds to DNA in the cell, it signals for the expression of certain genes to begin; this is accomplished in part by employing RNA polymerase.

56. Based on the pedigree chart below, what term best describes the nature of the trait being mapped?

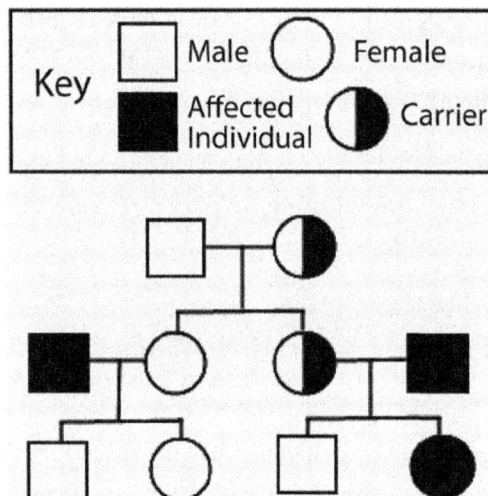

A. autosomal recessive

B. sex-linked

C. incomplete dominance

D. co-dominance

Answer: B.

Sex-linked. This chart would be a good example of color blindness, a sex-linked trait. If the trait had been autosomal recessive the last generation would all be carriers with the exception of the affected individual. In the case of traits that are incompletely dominant or co-dominant, the tree would require additional notation.

57. **Capillaries come into contact with a large surface of both the kidneys and lungs, especially in relation to the volume of these organs. Which of the following is not consistent with both organs and their contact with capillaries.**

A. Small specialized sections of each organ contact capillaries

B. A large branching system of tubes within the organ

C. A large source of blood that is quickly divided into capillaries

D. A sac that contains a capillary network within its interior volume

Answer: D.

A sac that contains a capillary network. The Bowmen's capsule of the kidneys can be described as a sac that contains a capillary network. The alveoli of the lungs are sacs, however the capillaries are on the outside of the alveoli.

58. **Samples of DNA were taken from a common species of tree frog in the temperate zones of North America. Samples came from individual frogs across the entirety of their range; however, upon analysis, researchers observed low genetic diversity in an isolated population found in the Ohio Valley. Based upon this observation, what would you hypothesize happened in the recent history of this population?**

A. A population bottleneck occurred with little to no migration into the valley.

B. Habitat loss and climate change introduced extrinsic forces that are selecting upon the population to adapt to a specialized niche.

C. Adaptive radiation for living in the Ohio Valley has resulted in this population.

D. The population experiences lower levels of genetic drift as a result of the small population.

Answer: A.

A population bottleneck occurred with little to no migration into the valley. Habitat loss and climate change could be the extrinsic factors that caused the bottleneck on the population, but without additional data on the population's history it would only be speculation. The bottelneck increases genetic drift within the population if there is no migration into the population. In small populations, infrequently occurring alleles face a greater chance of being lost, which can further decrease the gene pool. Due to the loss of genetic variation, the new population can become genetically distinct from the original population, which has led to the hypothesis that population bottlenecks can lead to the evolution of new species.

59. The term "vaccination" is etymologically related to the Latin word for cow, "vacca". Why?

A. The first species to be vaccinated was the cow, an animal of great economic importance in the 19th century. Veterinarians used an inactivated version of cowpox to protect cows against this deadly bovine disease.

B. Scientists originally investigated vaccines in the cow. Animal experiments in cows indicated that vaccines could be safely and effectively used in humans.

C. Cowpox is a similar virus to the serious human virus, smallpox, and was used as the original vaccine because it causes a mild infection in humans. The idea for this came from the observation that milkmaids did not contract smallpox.

D. When polio infects cows, it mutates into an attenuated form. When this attenuated virus is injected into humans, it causes a mild infection that serves as a vaccination against polio.

Answer: C.

Cowpox is a similar virus to the serious human virus, smallpox, and was used as the original vaccine because it causes a mild infection in humans. The idea for this came from the observation that milkmaids did not contract smallpox. The observation that milkmaids did not contract smallpox led Dr. Edward Jenner to test whether cowpox could protect humans from smallpox. The experimental animal for this was not a cow, but an 8-year-old boy! Luckily, the boy was protected by the vaccination and did not therefore die when he was subsequently injected with the deadly smallpox virus to test the idea! Interestingly, vaccination had its critics among the uninformed from day one—cartoons exist from the early experimental days of vaccination depicting people turning part-cow after vaccination.

60. In an aqueous solution, in which way do most of the hydrophobic side chains of amino acids orient themselves?

A. Exposed to the exterior of the protein to promote interactions with other molecules.

B. In a configuration that promotes their interaction with the peptide backbone.

C. Pointed inward such that they associate with one another.

D. In a random manner determined by the positioning of hydrophilic side chains.

Answer: C.

Pointed inward such that they associate with one another The majority of hydrophobic sidechains are oriented such that they form the "hydrophobic core" of the protein; this is largely driven by repulsive forces from the aqueous environment, which results in the more *hydrophilic* sidechains being exposed on the exterior of the protein. Note that formation of this hydrophobic core is often the first step in the derivation of a protein's tertiary structure.

61. Which of these would result from incapacitating the muscles of a mammalian eye?

A. The inability to regulate the amount of light entering the eye and an inability of the photoreceptor cells to transduce light energy

B. The inability of the photoreceptor cells to transduce light energy

C. The inability to focus light and an inability of the photoreceptor cells to transduce light energy

D. The inability to regulate the amount of light entering the eye and an inability to focus light

Answer: D.

The inability to regulate the amount of light entering the eye and an inability to focus light. The opening of the eye is called the iris and it controls the amount of light that can enter. If the muscles that control the iris are cut or damaged, the ability to regulate the amount of light entering the eye would be affected.

62. A sequence of pictures of polypeptide synthesis shows a ribosome holding two transfer RNAs. One tRNA has a polypeptide chain attached to it; the other tRNA has a single amino acid attached to it. What does the next picture show?

A. Both tRNAs remain, but the polypeptide chain is transferred to the single amino acid.

B. The tRNA connected to the single amino acid leaves the ribosome.

C. The tRNA connected to the polypeptide chain leaves the ribosome.

D. Both tRNAs remain, and a third tRNA with an amino acid joins the pair on the ribosome.

Answer: C.

The tRNA connected to the polypeptide chain leaves the ribosome. The chain-connected tRNA has already transferred its single amino acid to the growing polypeptide chain held in the ribosome. The next step is that this tRNA will leave, and as it does, the tRNA with the single amino acid will move into position to join its tRNA to the chain. Meanwhile, a new tRNA with a single amino acid will take the place just left open. However, at any given time, only two tRNAs will be in the ribosome—the one that has connected its amino acid to the growing chain and is about to leave, and the one that is still connected to an individual amino acid, that is waiting to move into position to connect its amino acid to the growing chain.

63. **The chemical reaction for photosynthesis is:**

$$6 \, CO_2 + 12 \, H_2O + \text{light energy} \rightarrow C_6H_{12}O_6 + 6O_2 + 6H_2O$$

A scientist used "heavy" water, with an oxygen with 10 neutrons instead of the normal 8 (^{18}O), as the water reactant, and finds that the O_2 gas released is composed of 100% ^{18}O. From these data, which of the following likely occurs during the reaction?

A. The CO_2 is split into 6 carbon atoms and 6 O_2 gas molecules, using NADPH

B. The CO_2 is split into 6 carbon atoms and 6 O_2 gas molecules, using light energy

C. The oxygen in the water molecules in the product come from the CO_2 molecules

D. The 12 H_2O molecules are split, and 6 oxygen atoms are incorporated into $C_6H_{12}O_6$ and 6 oxygen atoms are incorporated into H_2O

Answer: C.

The oxygen in the water molecules in the product comes from the CO_2 molecules. Since the input water contained ^{18}O, the presence of ^{18}O in the product oxygen gas indicates that the oxygen in the product water must be from the CO_2, the only "regular" oxygen among the reactants.

Grid-in:

1. A plant geneticist is investigating the inheritance of genes for sour taste (Su) and green skin (e) in apples (*Malus domestica*). Green skin is recessive and the sour taste are a result of the recessive genotype (susu). The geneticist wishes to determine if the genes assort independently and performs a testcross between a sour/red hybrid and a tree homozygous recessive for both traits. The following offspring are produced:

sour/red – 48

sour/green – 28

non-sour/red – 22

non-sour/green – 41

Calculate the chi-squared value for the null hypothesis that the two genes assort independently. Give your answer to the nearest tenth.

Answer: 5.7

Explanation: A cross between Susu/Ee x susu/ee apple trees is expected to produce offspring in a 1:1:1:1 phenotypic ratio. The chi-square value is calculated below. The range of acceptable answers for this question should be 5.6 to 5.7.

Phenotype	Observed	Expected	obs-exp	(obs-exp)2	(obs-exp)2/exp
sour/red	48	34.75	13.25	175.56	2.35
sour/green	28	34.75	-6.75	45.56	0.61
non-sour/red	22	34.75	-12.75	162.56	2.17
non-sour/green	41	34.75	6.25	39.06	0.52
					X^2=5.65

2. In fruit flies, Drosophila, a phenotype of eyes that look like small slits or kidney shaped "bar eye" is a result of a mutation displaying an autosomal dominant pattern of inheritance. Homozygous recessive individuals (B^+B^+) display a wild-type phenotype with normal eyes. Inheriting two copies of the mutation (BB) is lethal during embryonic development. In a cross between strains of drosophila with the bar eye phenotype, what proportion of eggs laid is expected to be bar eyed? Give your answer in the form of a fraction.

Answer: 2/3.

Explanation: The homozygous dominant flies died out, leaving 1/3 of the remaining flies homozygous recessive and 2/3 heterozygous.

3. The following are lengths of bull head minnows, measured from nose to tail, recorded in centimeters. Calculate the mean. Give the answer in cm to the nearest tenth.

15.4   10.2   14.6   9.1   9.2   13.1   12.5   15.2   11.2

Answer: 12.3.

Explanation: Calculate the mean by adding the set and dividing by the number of datapoints. The result is 12.3.

4. Calculate the standard deviation for the data set given in the previous question. Give the answer in cm to the nearest tenth.

Answer: 2.5.

Explanation: Calculate the standard deviation for the above set. The result is 2.5.

5. In peas (*Pisum sativum*), wrinkled peas (W) are dominant to smooth peas (w). Peas from the offspring of a cross between wrinkled-pea and smooth-pea plants were used in a lab. A student counts 429 wrinkled and 399 smooth peas in the pod. Calculate the chi-squared value for the null hypothesis that the purple parent was heterozygous for wrinkled peas. Give your answer to the nearest tenth.

Chi-Square Table								
Degrees of Freedom								
p	1	2	3	4	5	6	7	8
0.05	3.84	5.99	7.82	9.49	11.07	12.59	14.07	15.51
0.01	6.64	9.32	11.34	13.28	15.09	16.81	18.48	20.09

Answer: 1.4

Explanation: A cross between a heterozygous wrinkle pea plant (Ww) and a smooth plant (ww) would yield offspring that display a 1:1 ratio between wrinkle and smooth peas. Of the 828 peas, it would be expected that 414 would be wrinkled and 414 would be smooth. The chi-square value is calculated below. The acceptable answer for this question should be 1.4

Phenotype	Observed	Expected	obs-exp	(obs-exp)2	(obs-exp)2/ exp
Wrinkle	429	414	15	225	0.72
Smooth	399	414	-15	225	0.72
					X^2=1.44

6. A tropical plant displays red flower color or pink flower color, with red the dominant phenotype, in an autosomal dominant pattern of inheritance. In a cross between two heterozygous strains of the flower, what proportion of offspring plants will have pink flowers? Give your answer in the form of a fraction.

Answer: 1/4.

Explanation: Pink flowers are recessive. This is a classic Mendelian cross; the offspring will display 25% homozygous recessive offspring.

Free Response: _____

1. **Describe the process of signal transduction, including an example.**

   Model Answer:

   Signal transduction begins when the ligand, a chemical signal that could be a peptide or a protein, is recognized by its receptor. After the ligand binds its receptor, the shape of the protein changes and allows the signal to be converted to a response by the cell. Possible examples: G-protein linked receptors – bind the energy-rich protein GTP **OR** ion channels – play an important role in the nervous system, as neurotransmitters released at a synapse allows the channel to open and propagate the signal **OR** Receptor tyrosine kinases – cell membrane receptors that have enzymatic activity through kinases that can phosphorylate proteins, resulting in their activation. This results in a signaling cascade to promote or inhibit target gene expression.

2. **The Endosymbiotic Theory proposes that mitochondria and chloroplasts originate through symbiosis. <u>Describe</u> the characteristics of these organelles that led scientists to hypothesize that they were once free-living organisms.**

   Model answer:

   The mitochondria and chloroplasts of present day eukaryotes resemble bacteria in many aspects of their structure and metabolism. For instance, mitochondria, chloroplasts and bacteria have a single circular DNA molecule, while the nuclear DNA is linear and arranged in multiple chromosomes. New mitochondria and plastids are formed inside the cells in a similar way to binary fission, which is the division mode of bacterial cells. Also, the organelles' ribosomes are 70S, as in bacteria, while cytoplasmic ribosomes are 80S. Some chloroplasts have a peptidoglycan cell wall in the membrane.

3. **Name three differences between plants and fungi.**

   Model answer:

   Fungi are heterotrophs; plants are autotrophs. Fungi reproduce with a dominant haploid generation; plants reproduce mainly by alternation of generations. Fungi do not necessarily have tissue differentiation; tissue differentiation is essential in plants as they need one tissue to photosynthesize and one to obtain water and minerals, at the simplest level.

4. **What are the three steps involved in the structural changes that occur during the mitotic process?**

   Model answer:

   Three distinct stages occur in the changes to the structure of the cell during mitosis. 1) Replication, the doubling of the chromosomes and the crossover; 2) alignment, when the nucleus dissolves, microtubules attach to the centromeres, and the chromosomes will align at the middle of the cell; and 3) separation, beginning with the pulling apart of the chromosomes, the disappearance of the microtubules, and cytokinesis resulting in two identical cells.

5. **Explain how stabilizing selection works in an ecosystem. Provide an example.**

Model Answer:

Stabilizing selection works to eliminate those forms of the population that fall at the extreme ends. It favors those with a more intermediate form of the trait. Dogs that are born with a very high or a very low birth weight do not tend to survive as long as those born with a average weight for the species.

6. **Describe the process of ecological succession after a volcanic explosion.**

Model Answer:

After the explosion the area would be devoid of life. Once the temperature cooled and conditions became favorable, lichens would move in and attach themselves to rocks. The acids produced by the lichens break down the rocks into sand and minerals. As the lichens die, their decaying bodies add nutrients to the sand turning it into soil. Once enough nutrients have accumulated, small grasses will move in. As these die and add more nutrients to the soil, larger plants will arrive. At the same time insects and other animals that feed on plants will inhabit the area. Over time, the small plants will be outcompeted by larger ones until large trees are the dominant species.

7. **When plants moved onto land, they needed to evolve several structures that allowed them to survive in an environment with much less water. These adaptations have afforded plants much success on land.**

   (A) Briefly **discuss** the evolution of land plants from their aquatic ancestor. Include from where plants are believed to have evolved.

   (B) **Describe** three structural adaptations and functions that plants evolved in order to live on land.

   (C) **Explain** the relationship that evolved between plants and nitrogen-fixing symbiotic bacteria in the soil.

Model Answer:

(A) The aquatic ancestors of land plants are the algae. Algae had already evolved into eukaryotic organisms through the symbiosis of photosynthetic bacteria (cyanobacteria) engulfed by cells probably related to the Archaea. At this point in evolution, the eukaryotic algae were capable of photosynthesis and more complex functions unique to eukaryotes. In addition, many algae had evolved to become multicellular, with different types of cells having different functions. For example, some cells were gametes and other cells were parts of the "leaves" (lamina) of seaweed, functioning in non-reproductive roles. However, the same tissues that could obtain energy through photosynthesis could also obtain minerals and water, simply by absorbing them from the surrounding seawater matrix. When plants moved onto land, they had to separate these functions-- plant tissues that absorbed water and nutrients had to be inside the earth, where light did not reach, and tissues involved in photosynthesis were not in contact with minerals and in limited contact with water that could be absorbed (limited to rain, dew, etc.), so plants needed to evolve structures such as roots and true leaves. Early land plants were less specialized than later ones, with leaf-like structures that lay flat on the ground, whereas later land plants evolved highly specialized structures that allowed survival in diverse dry-land environments.

(B) Plants evolved to have two fundamental structural adaptations in order to live on land, roots and leaves. A third fundamental structural adaptation is a vascular system that connects these two types of tissues, allowing the delivery of minerals and water from the roots to the leaves and the delivery of sugars from the leaves to the roots. (Nonvascular plants such as liverworts do not have specialized vascular tissues like xylem and phloem, so they have a flat shape, with leaf-like tissues right next to the ground. These rudimentary leaves also do not have many of the structures of higher plant leaves that allow plants to resist desiccation-- this allows the liverworts to absorb water but also makes them vulnerable to drying out, so they are limited to moist environments.) Another adaptation that allowed plants to live in dry environments is the protective reproductive structures such as coated spores and seeds. However, not all plants have such structures; mosses, for example, have swimming gametes and therefore can only reproduce in moist environments that allow their gametes to swim to each other. Correct answers for this question could include a number of specific structures, such as stomata on leaves that can close to hold in water, protective walls on spores that allow them to tolerate dry conditions, waxy cuticles to reduce water loss in leaves, roots that store water in dry environments, and protective structures around embryos-- most notably, seeds. As long as you can describe their function, any number of specific structures could fulfill this question. If you encounter a similar question on the AP exam, you may want to mention several structures and functions-- brainstorm a bit-- since points are not deducted on the Free Response section for extra information. Just be sure to keep an eye on the clock and stay within your allotted time for the question.

(C) Nitrogen is just one of the compounds that plants must obtain from the soil. 79% of the air in the atmosphere is nitrogen, so it is a very plentiful element, but this nitrogen is in the form of N_2 gas, which plants cannot use. Nitrogen gas can be "fixed", or made into the more usable form of NH_4, by bacteria in the soil. Some plants, such as the legumes, have evolved a system of symbiosis with nitrogen-fixing bacteria. Legumes have root nodules filled with nitrogen fixing bacteria called *Rhizobium*, giving the bacteria a protective environment and nutrients, while in return, the plant receives usable nitrogen in the form of ammonia. While the symbiosis between legumes and *Rhizobium* is the most well-known, many other bacterial genera fix nitrogen, and some of these form symbioses with various plants. The fact that different types of nitrogen-fixing bacteria have formed symbiotic relationships with different types of plants is evidence for the importance of nitrogen and the nitrogen cycle in plant survival.

8. **Describe the materials or factors involved in the photosynthetic reaction.**

Model Answer:

- Light energy: The radiant energy from the sun is a source of light energy for photosynthesis. Light energy is harvested by the pigments in order to break down water molecules into hydrogen and oxygen.

- Carbon Dioxide: During photosynthesis, carbon dioxide is converted into sugars for energy. The carbon dioxide of the atmosphere is used by plants.

- Water: During photosynthesis, water is absorbed by plant roots and through photolysis, oxygen is released and the hydrogen is used to fix carbon dioxide.

- Chlorophyll: Pigments capable of absorbing energy from the sun. There are two types of photosynthetic pigments - chlorophylls and carotenoids. Chlorophylls are the main pigments as they are involved in the conversion of light energy into chemical energy, while carotenoids absorb light energy and pass it to the chlorophyll molecules.

1. During pregnancy, the body undergoes a number of changes to facilitate the birth of the baby. Among these are changes that allow the bones of the pelvis to be pulled apart to widen the birth canal. Which of the following structures would be expected to relax to accomplish this?

    A. Ligaments

    B. Tendons

    C. Muscles

    D. Bones

2. *Dalbulus maidis* is an insect pest of domesticated maize (corn) and its wild ancestor, teosinte, both of which originated in central Mexico. *D. maidis* eggs are laid on the midrib and leaf blade in teosinte but only on the midrib of the upper leaf surface in maize. A Mexican biologist observed that larvae of another pest, the chewing insect *Spodoptera frugiperda*, consume the leaf blade of teosintes and maize, but not the midrib, and he thought that this might explain the egg-laying behavior of *D. maidis*. Which of the following could explain his conclusion?

    A. Maize, planted in dense fields, hosts larger populations of pests. In the context of a large population of *S. frugiperda*, the *D. maidis* eggs are at risk of being eaten by *S. frugiperda* if they are laid on the blade. This selection pressure has led to niche partition between the two species.

    B. Teosinte, in its diverse natural habitat, hosts a smaller population of pests. This provides plenty of food for all of the insects, and this selection pressure has led to niche specialization by the two species.

    C. As it has evolved into a different species from its ancestor teosinte, maize has evolved chemical defenses to repel insect pests. The selection pressure of the different insects has led to specialized defenses that are toxic to *D. maidis* on the midrib and *S. frugiperda* on the leaf blade.

    D. Maize has a much more diversified herbivore population, including *D. maidis* and *S. frugiperda*, since it is more genetically heterogeneous and evolutionarily basal to teosinte. The dependence of maize on insect herbivores has given rise to selection pressure that has led to niche partition between the two species.

3. Urinary tract infections start when bacteria outside the body enter the urinary system and move up into the body. Choose the correct order of structures in women that might be invaded and lead to infection.

A. The urethra, then the bladder, then the ureter, eventually leading to a liver infection

B. The vagina, then the bladder, then the urethra, eventually leading to a kidney infection

C. The urethra, then the bladder, then the ureter, eventually leading to a kidney infection

D. The vagina, then the urethra, then the ureter, eventually leading to a bladder infection

For questions 4-7, refer to the following phylogram. This phylogenetic tree is based on the reverse transcriptase amino acid sequence of several LTR retrotransposons, viral-like mobile genetic elements. These types of retrotransposons are similar to retroviruses, except that they lack the viral envelope needed to be transmitted from one host organism to another (to be infectious). They are also known as endogenous retroviruses.

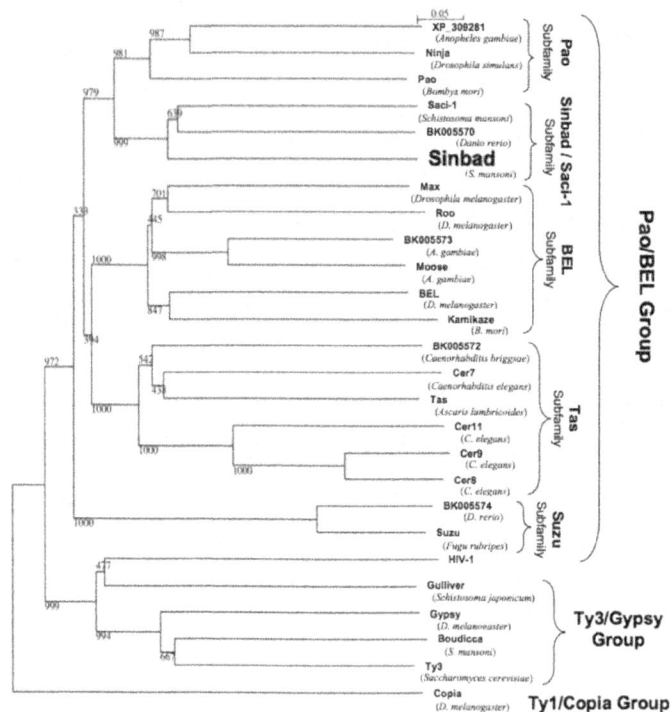

4. In the phylogram above, which of the following is the most evolutionarily distant from the Ty3 element?

A Pao

B. HIV-1

C. Copia

D. Sinbad

5. Which of the following is most closely related to the *Boudicca retrotransposon*, a retrotransposon from the African blood fluke *Schistosoma mansoni*?

    A. HIV-1, from *Homo sapiens*

    B. Copia, from *Drosophila melanogaster*

    C. Cer7, from *Caenorhabditis elegans*

    D. Sinbad, from *Schistosoma mansoni*

6. The closest relative of the Sinbad retrotransposon from *Schistosoma mansoni*, a parasitic flatworm, is very close on the phylogram to a sequence from *Danio rerio*, the zebrafish. Which of the following is one possible explanation for this?

    A. *Schistosoma mansoni* is very closely related to *Danio rerio*.

    B. The flatworms, phylum Platyhelminthes, evolved from fish.

    C. *Schistosoma mansoni* and *Danio rerio* both evolved Sinbad-like retrotransposons independently; this is known as convergent evolution.

    D. A Sinbad-like retrotransposon may have been horizontally transmitted from an ancestor of *S. mansoni* to an ancestor of *Danio rerio*.

7. *Anopheles gambiae* is a mosquito, *Drosophila* spp. are fruit flies, *Bombyx mori* is the silkworm (a moth), *Schistosoma* spp. are parasitic flatworms, *Danio rerio* and *Fugu rubripes* are fish, *Ascaris* is a parasitic roundworm, and *Caenorhabditis* spp. are free-living roundworms. Which of the Pao/BEL group subfamilies include retrotransposons that are most likely to have undergone horizontal transmission?

    A. Pao

    B. Sinbad

    C. Tas

    D. Suzu

8. If a person has gallbladder cancer, on the basis of the pain that they feel, what organ might a doctor mistakenly think is diseased?

    A. The kidneys

    B. The liver

    C. The appendix

    D. The heart

9. **The wild relatives of maize (corn) have trichomes on their early leaves, whereas commercial maize has lost this trait. Recent research from the University of Guadalajara suggests that this early pubescence my serve as a physical barrier to thwart insect pest herbivores and protect the young plants. What does the word "pubescence" mean in this context?**

   A. Adolescence

   B. Being hairy

   C. Being mature enough to produce seeds

   D. Producing a natural insecticide

10. **Which type of snack would require the most amylase activity for digestion?**

    A. Hard candy

    B. Dried meat (jerky)

    C. Toast

    D. Butter

11. **In vascular plants, the dominant part of the plant is diploid. Which of the following statements about vascular plants are true?**

    A. The sporophytes of vascular plants create gametes by meiosis.

    B. The sporophytes of vascular plants create gametes by mitosis.

    C. The sporophytes of vascular plants create spores by meiosis.

    D. The sporophytes of vascular plants create spores by mitosis.

12. **Human kids are well-known to hate vegetables and spicy food like chili peppers, but adult humans often love these flavors. How can this be explained, in terms of physiology and evolution?**

    A. Plants have evolved compounds during phylogeny to repel herbivores but the human body learns during ontogeny that the compounds are healthy, leading to a preference for the flavors.

    B. Plants have evolved compounds during ontogeny to repel herbivores but the human body learns during phylogeny that the compounds are healthy, leading to a preference for the flavors.

    C. Plants have evolved compounds during phylogeny that enhance human health, but these compounds are not healthy for children.

    D. Children have evolved a repulsion for bitter compounds during phylogeny but plants have evolved compounds that appeal to adults during ontogeny.

13. Kidney stones form through the precipitation of dissolved calcium and oxalate in the kidney under acidic conditions. A high animal protein intake significantly increases excretion of calcium, oxalate, and uric acid into the urine. Which of the following will provide the BEST reduction of kidney stones?

A. A low-meat diet and a high fluid intake

B. A vegetarian diet and a low fluid intake

C. A high-meat diet and a high fluid intake

D. A vegan (no meat or dairy products) diet and a high fluid intake

14. Researchers at the University of Guadalajara found that two species of teosinte, the wild ancestor of corn, expressed trichomes at an early developmental stage whereas modern commercial maize (corn) did not express this phenotype until a later developmental stage. The trichomes, fine hairs growing out of the leaf blades, prevented chewing by the generalist pest insect *Spodoptera frugiperda*, resulting in significantly less damage in the teosintes than in the modern maize. Which of the following is most likely to be true?

A. The two wild species independently developed the early-trichome trait, since both were under selection pressure from the herbivore, *Spodoptera frugiperda*.

B. One wild species developed the early-trichome trait under pressure from the herbivore, and passed it to the other wild species by conjugation.

C. The wild species from which modern maize developed had the early-trichome trait but modern maize lost it because pesticide use reduced selection pressure for the trait.

D. Modern maize had the early trichome trait, then passed it to one of the wild species through a transposon.

15. Considering the anatomy of the human body, if a kidney stone is large enough to fill the renal pelvis, which of the following is true?

A. The stone can be passed by giving the patient a large amount of fluid.

B. The stone is too large to be passed; it will cause serious pain and eventually gallbladder dysfunction.

C. The stone must be disrupted (e.g. by sound waves) or removed surgically.

D. An alpha blocker should be administered to relax the muscles of the ureter and then the stone can be passed.

For questions 16-20 refer to the drawing and description below:

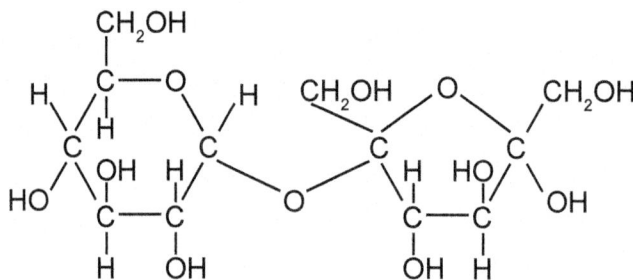

Under anaerobic conditions, yeast will ferment this compound to produce energy. However, before this can happen, they must break this disaccharide into two monosaccharides, glucose and fructose, both of which have the same molecular formula: $C_6H_{12}O_6$. The unbalanced equation of glucose or fructose fermentation is:

$$C_6H_{12}O_6 \rightarrow C_2H_5OH + CO_2$$

16. In order to break this disaccharide into two monosaccharides, what type of small molecule will participate in the reaction, adding its atoms to the final two monosaccharides?

    A. CO_2

    B. H_2O

    C. $C_6H_{12}O_6$

    D. CH_2COOH

17. What enzyme might catalyze the metabolism of sucrose into glucose and fructose?

    A. Glycoside hydrolase

    B. Protease

    C. Lipase

    D. Chantililase

18. When a winemaker makes wine using yeast fermentation, what are the products of the fermentation of a disaccharide like that pictured above?

    A. Fructose and glucose

    B. Ethanol and water

    C. Ethanol and carbon dioxide

    D. Fructose and ethanol

19. **How would the energy yield of anaerobic fermentation compare with the energy yield from the same disaccharide under aerobic conditions?**

   A.   Fermentation yields much more energy.

   B.   Fermentation yields much less energy.

   C.   Fermentation yields the same amount of energy.

   D.   Fermentation yields a little less energy.

20. **If the yeast were given the same disaccharide under aerobic conditions, which of the following would be involved in the metabolism that would take place under these conditions?**

   A.   The Calvin cycle, mitochondria, electron transport chain

   B.   The Krebs cycle, mitochondria, chemiosmotic potential

   C.   The Krebs cycle, electron transport chain, carbon fixation

   D.   The Krebs cycle, electron transport chain, fermentation,

21. **Can human bodies harvest energy from the ethanol in wine?**

   A.   No, ethanol cannot be fermented because it is already the product of fermentation.

   B.   No, ethanol can only be detoxified using alcohol dehydrogenase.

   C.   Yes, ethanol can be processed and then enter the Krebs cycle and be metabolized to CO_2.

   D.   Yes, ethanol can be fermented via anaerobic respiration.

22. **Two closely related crop plants produce a natural flavor compound that is appealing to humans and also detectable by a common insect pest. One subspecies (Plant A) begins producing the flavor compound at the 2-leaf stage of development. The other subspecies (Plant B) does not begin producing it until the 8-leaf stage, but produces much more of it than Plant A. Plant B requires substantially more chemical pesticide than Plant A to reduce damage from this insect to produce a good crop yield. What does this say about the feeding preferences of the insect?**

   A.   The compound is toxic to the insect, which prefers to eat older plants.

   B.   The insect likes the flavor of the compound, and prefers to eat younger plants.

   C.   The insect is repelled by the compound, and prefers to eat older plants.

   D.   The insect is repelled by the compound, and prefers to eat younger plants.

23. Textbooks often say that respiration yields 38 molecules of ATP, but the real yield is closer to 30 molecules of ATP. What might account for this?

A. Leaky membranes

B. Non-metabolic mitochondria

C. Partial anaerobic metabolism/fermentation

D. Some of the ATP is used to power the Krebs cycle

Questions 24-27. Parasitoid wasps can be useful in natural pest control, since many are natural predators of crop pest insects. A team of researchers from the USDA and the Centro de Ecologia in Mexico, looking for an all-female strain of parasitoids, found a strain of the fruit fly parasitoid *Odontosema anastrephae* that included only females, with no males after several generations. The following figure is a phylogenetic tree of the cytochrome C oxidase subunit 1 (COI) gene of individuals of the parthenogenetic (all-female) and arrhenotokous (male and female) strains. The numbers on the branch node are "bootstrap values"—out of 1,000 repetitions of the tree-drawing process, this is how many times the computer has given that particular tree configuration.

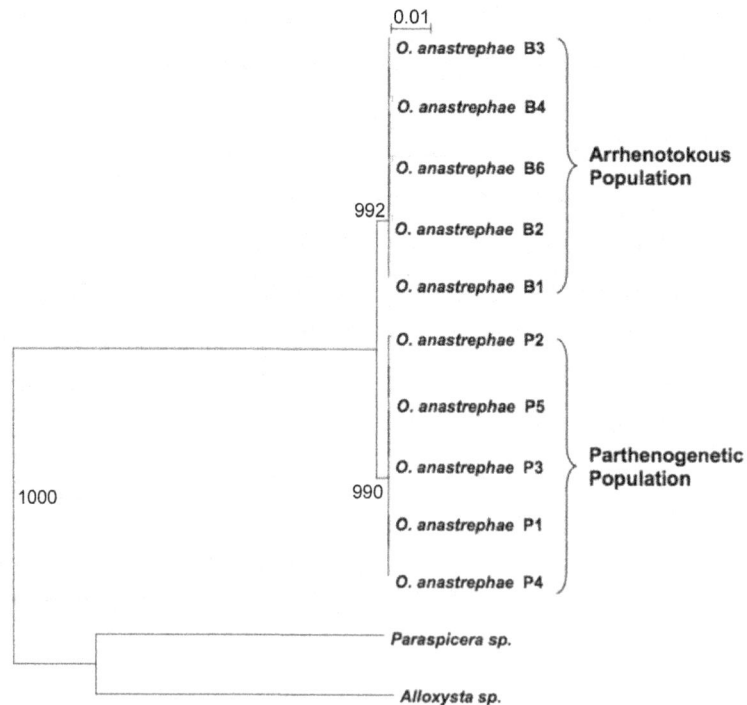

24. **Based on the phylogram above, what could you conclude about these wasps?**

   A. The wasps represent two populations that are closely related but genetically distinct; they probably will not interbreed if given the opportunity.

   B. There is no significant difference between the two populations; they will probably interbreed if they are given the opportunity.

   C. The two populations are genetically distinct; they actually are different species. They will definitely interbreed if given the opportunity.

   D. The parthenogenetic wasps are a subspecies of the arrhenotokous wasps, but this branch has recently diverged so they will probably interbreed if given the opportunity.

25. **Considering the reproductive strategy of these wasps, what must be true of the parthenogenetic *O. anastrephae*?**

   A. In the parthenogenetic strain, mating-type alpha female wasps reproduce sexually with mating-type alpha females

   B. Female wasps of the parthenogenetic strain only have one parent, but will nevertheless be unique because of the effect of crossing over

   C. Wasps of the parthenogenetic strain are clones – either full or half-clones

   D. Male wasps of parthenogenetic strain are clones of their mother

26. **USDA biologists were looking for all-female wasps because it would cut production costs of these natural predators in half. Knowing this, what is probably true about these predators?**

   A. Both male and female wasps eat many insect larvae per day, but females eat more than males.

   B. Male wasps are more expensive to produce because they are larger.

   C. The wasps parasitize the pest insects by burrowing into the larvae.

   D. The wasps parasitize the pest insects by laying eggs inside the larvae.

27. **The cytochrome C oxidase subunit I gene is often used in phylogenetic comparisons like the one in the previous questions. Why?**

    A. Cytochrome C functions as part of the immune system, and is therefore highly conserved in all eukaryotes.

    B. Cytochrome C functions as part of the electron transport chain, and is therefore highly conserved in all eukaryotes.

    C. Cytochrome C functions as part of the Calvin cycle, and is therefore highly conserved in all eukaryotes.

    D. Cytochrome C is only found in insects and is therefore useful for distinguishing between subspecies.

28. **Which of the following statements are true?**

    A. Fungi are to slime molds as plants are to algae.

    B. Plants are to algae as cyanobacteria are to mitochondria.

    C. Cyanobacteria are to chloroplasts as eubacteria are to mitochondria.

    D. Archaea are to mitochondria as cyanobacteria are to chloroplasts.

29. **Plasmodium is a protozoan parasite that causes malaria. Which of the following is true about Plasmodium?**

    A. It is a multicellular member of the kingdom Animalia in the older system of classification, and domain Archaea in the modern three-domain system of classification.

    B. It is a multicellular member of the kingdom Protista in the older system of classification, and domain Eubacteria in the modern three-domain system of classification.

    C. It is a unicellular member of the kingdom Monera in the older system of classification, and domain Eubacteria in the modern three-domain system of classification.

    D. It is a unicellular member of the kingdom Protista in the older system of classification, and domain Eukarya in the modern three-domain system of classification.

30-33. Luciferase is an enzyme that produces light in fireflies. The amount of luciferase expressed in a cell can be deduced by measuring its light output in a luminometer. A biologist was interested in knowing whether the promotors for two retrotransposons she had discovered, Boudicca and Sinbad, were functional. Molecular biologists make use of plasmids, small circular DNAs that are naturally exchanged between bacteria, to look at gene expression. Stretches of DNA the researcher is interested in studying are inserted into the plasmid, and the biologists use lab techniques to get bacteria to take up the plasmids. Questions 30-33 refer to the following figure:

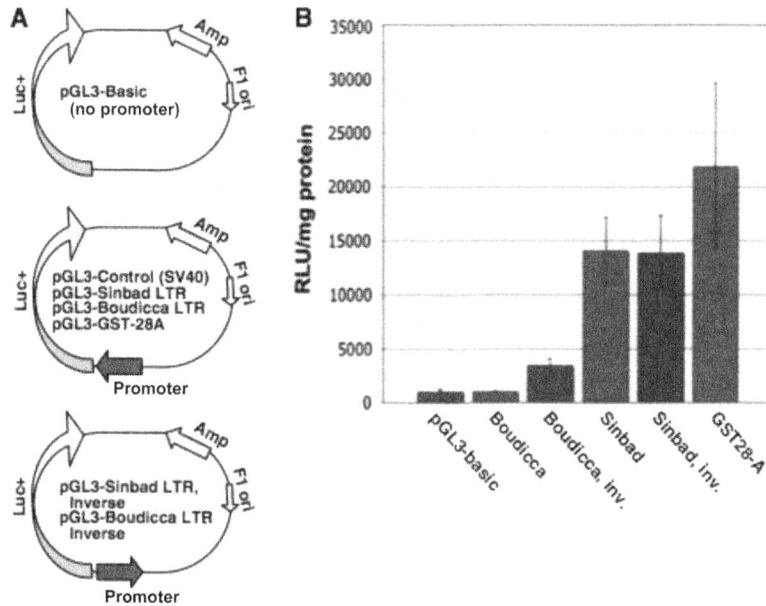

30. Part A of the figure shows lab-created plasmids (called expression vectors), made from inserting genes or other genetic elements into a bacterial plasmid. These plasmids were constructed to test whether a gene promotor is functional or not. Recalling the structure of the *Lac* operon, which of the two vectors are negative controls?

A. The top two vectors

B. The bottom two vectors

C. The top and bottom vectors

D. All of them are controls

31. **Part B of the figure shows light output measured in a luminometer. The GST-28A promoter is known to be functional. According to the graph of light readings, which other promoter is functional?**

A. The Boudicca promoter

B. The Sinbad promoter

C. Both the Boudicca and the Sinbad promoter

D. Neither the Sinbad nor the Boudicca promoter

32. **Part B of the figure shows a surprising, unexpected result. What is this result and what does it indicate?**

A. The Boudicca promoters were nonfunctional. This indicates that the Boudicca retrotransposon does not encode luciferase.

B. The GST-28A promoter had a very large error bar. This indicates that this promoter was not a good choice for a positive control.

C. The Sinbad promoter did not express as much luciferase as the GST-28A promoter. This indicates that the Sinbad promoter is probably nonfunctional.

D. Both the forward and inverted Sinbad promoters were functional. This indicates that this is an unusual class of promoter, known as a bidirectional promoter.

33. **In Part A of the figure, part of the vector plasmid is labelled "amp" for ampicillin resistance—this is where an ampicillin resistance gene has been cloned in. Why has an ampicillin resistance gene been cloned into the vector?**

A. So that the researcher can know whether bacteria have been successfully transformed with the plasmid.

B. So that the researcher can make sure the organisms with the plasmid do not become infected with bacteria.

C. So that the plasmids can be used in animals being prophylactically treated with antibiotics.

D. So that the plasmid is protected against ampicillin-resistant bacteria that would interfere with the experiment.

Questions 34-41. A team of researchers from the USDA and the Centro de Ecologia in Mexico discovered a strain of parasitic wasps that was all-female. These wasps are natural predators of fruit flies, a natural alternative to chemical pesticides. A genetic study indicated that, while the all-female and male and female (bisexual) strain were morphologically similar, the all-female strain may in fact be a different species. After observing that the females of the all-female strain did not mate with the males of the bisexual strain, they did a series of experiments to see whether the behavior of the two strains was statistically different. In the experiments, they used an olfactometer, a multi-armed chamber in which one arm contained bait such as a guava infested with the fruit fly prey of the wasps. They then measured the time spent in the target arm of the device. Questions 34-41 refer to the figure to the right, which describes these experiments. Thelytokous is another name for an all-female strain and arrhenotokous is another name for the bisexual strain.

34. In this experiment, what is the dependent variable and what is the independent variable?

   A. The independent variable is the time spent in the target arm of the olfactometer; the dependent variable is the strain of wasp (all-female or bisexual).

   B. The independent variable is the strain of wasp (all-female or bisexual); the dependent variable is the time spent in the target arm of the olfactometer.

   C. The independent variable is which arm of the olfactometer the wasp went to; the dependent variable is the strain of wasp (all-female or bisexual).

   D. The independent variable is the strain of wasp (all-female or bisexual); the dependent variable is which arm of the olfactometer the wasp went to.

35. Part A represents the time it took the wasps to find the guava bait. Infested guavas were tested in one experiment, and non-infested guavas were compared in a separate experiment. Which of the following statements describes the results?

   A. The all-female wasps found the non-infested guava significantly faster than the bisexual wasps.

   B. The bisexual wasps found the non-infested guava significantly faster than the all-female wasps.

   C. There was no statistical difference in time needed to find the non-infested guava.

   D. Both strains of wasps took longer to find the infested guava than the non-infested guava.

36. **Part B represents the time spent in the arm with the target guava. Infested guavas were tested in one experiment, and non-infested guavas were compared in a separate experiment. Which of the following statements describes the results?**

A. The arrhenotokous wasps spent more time in the target arm than the thelytokous wasps, for both infested and non-infested guavas.

B. The thelytokous wasps spent more time in the target arm than the arrhenotokous wasps, for both infested and non-infested guavas.

C. The arrhenotokous wasps spent significantly more time with the infested guava than the thelytokous wasps, but the thelytokous wasps spend more time with the non-infested guava.

D. The thelytokous wasps spent significantly more time with the non-infested guava than the arrhenotokous wasps, but the time spent with the infested guava was not statistically different between the two.

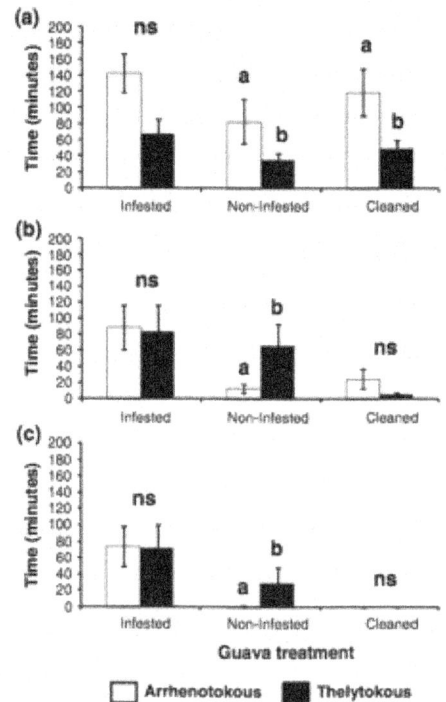

37. **This type of wasp finds its prey by entering the fruit and swimming through the pulp until it finds a fruit fly larva to parasitize. Part C shows the amount of time spent by the wasps actually swimming in the guava in the target arm. What conclusions can be drawn from the data shown in parts B and C?**

A. The thelytokous wasps behave similarly to the arrhenotokous wasps when prey is present, but spend significantly less time exploring non-infested fruit, which could be a fitness disadvantage.

B. The thelytokous wasps behave similarly to the arrhenotokous wasps when prey is present, but waste more time on non-infested fruit, which could be a fitness disadvantage.

C. The thelytokous wasps prefer non-infested guavas, while the arrhenotokous wasps prefer infested guavas, which could be a fitness advantage for the arrhenotokous wasps.

D. The thelytokous wasps prefer non-infested guavas, while the arrhenotokous wasps prefer infested guavas, which could be a fitness disadvantage for the arrhenotokous wasps.

38. **Which of the following represents a reasonable speculation regarding these results—something that cannot be concluded but would be a reasonable hypothesis for further research?**

A. Perhaps the thelytokous wasps are good at smelling guava, but are not very good at smelling fruit fly larvae (thelytokous wasps orient to guava).

B. Perhaps the arrhenotokous wasps are good at smelling fruit fly larvae, but not very good at smelling guava (arrhenotokous wasps orient to fruit fly larvae).

C. Perhaps the arrhenotokous wasps use smell to detect prey, whereas thelytokous wasps use sight (arrenhotokous wasps orient by smell; thelytokous wasps orient by vision).

D. Both A and B.

39. **What experiments could test the hypothesis that the arrhenotokous wasp strain orients to fruit fly larvae whereas the thelytokous wasp strain orients to fruit?**

A. (1) Use the same olfactometer set-up, but with larvae only instead of guava; (2) Use an olfactometer with a non-infested guava in one arm and larvae in another arm.

B. (1) Use the same olfactometer set-up, but with a different fruit infested with the fruit fly larvae; (2) Use an olfactometer with this different fruit in one arm and an infested guava in another arm

C. (1) Use the same olfactometer set-up, but with guavas infested with dead larvae; (2) Use an olfactometer with a guava infested with live larvae in one arm and dead larvae in another arm

D. (1) Use the same olfactometer set-up, but with guava leaves instead of guava fruit; (2) Use an olfactometer with an infested guava in one arm and a mammal in another arm

40. **What trophic level do these wasps represent in a cloud forest ecological setting?**

A. The first trophic level

B. The second trophic level

C. The third trophic level

D. The fourth trophic level

41. **These two wasp strains may or may not be the same species. However, knowing that they are parasitoid wasps, what higher taxonomic category are they in and what category includes the evolutionary ancestor of both the wasps and their fruit fly prey?**

A. The wasps are in the order Hymenoptera and both the wasps and the fruit flies are in the class Insecta (phylum Arthropoda).

B. The wasps are in the phylum Hymenoptera and both the wasps and the fruit flies are in the order Arthropoda (class Insecta).

C. The wasps are in the order Arthropoda and both the wasps and the fruit flies are in the class Hymenoptera (phylum Insecta).

D. The wasps are in the class Arthropoda and both the wasps and the fruit flies are in the order Insecta (phylum Hymenoptera).

42. **Which of the following is true about the reproduction of fungi?**

A. Most fungi consist of diploid hyphae until it is time for sexual reproduction. At this point, hyphae from two fungal organisms fuse into a dikaryotic organism (two nuclei per cell), which forms sexual sporangia in which the nuclei fuse into one diploid nucleus, which then undergoes meiosis to form diploid spores.

B. Most fungi consist of haploid hyphae until it is time for sexual reproduction. At this point, hyphae from two fungal organisms fuse into a dikaryotic organism (two nuclei per cell), which forms sexual sporangia in which the nuclei fuse into one diploid nucleus, which then undergoes meiosis to form haploid spores.

C. Most fungi consist of haploid hyphae until it is time for sexual reproduction. At this point, hyphae from two fungal organisms fuse into a diploid organism (two nuclei per cell), which forms sexual sporangia which then undergo meiosis to form diploid spores.

D. Most fungi consist of diploid hyphae until it is time for sexual reproduction. At this point, hyphae from two fungal organisms fuse into a haploid organism, which forms sexual sporangia in which the nuclei fuse into one haploid nucleus, which then undergoes meiosis to form diploid spores.

43. **Which of the following statements about the ecology of fungi is true?**

A. Fungi are autotrophs; most are saprophytes and some are photosynthetic.

B. Fungi are heterotrophs; most are herbivores and some are predators.

C. Fungi are heterotrophs; most are parasites and some are predators.

D. Fungi are heterotrophs; most are saprophytes and some are parasites.

44. In the traditional view of the ability of plants to colonize land, factors such as tissue differentiation have been considered. Another factor that many biologists now consider essential for the transition of plants to land was association of plants with fungi. What is the general name for this type of biological partnership, and what is the specific name for the structures that form from the relationship between soil-dwelling fungi and plants?

   A. Parasitism; mycorrhizae

   B. Symbiosis; fungorrhizomes

   C. Symbiosis; mycorrhizae

   D. Parasitism; fungorrhizomes

45. Which of the following describes the ecological relationship between protozoans and algae?

   A. Algae are consumers and protozoans are producers; together, they provide food for other consumers in the food web.

   B. Algae are producers and protozoans are consumers; together, they provide food for other consumers in the food web.

   C. Algae are consumers and protozoans are producers; together, they provide food for other producers in the food web.

   D. Algae are producers and protozoans are consumers; together, they provide food for other producers in the food web.

46. Diego and Darius are eating lunch. Diego is eating sushi, which contains seaweed, rice, and salmon. Darius is eating a grilled blue cheese and apple sandwich on whole wheat bread. At what trophic levels are Diego and Darius?

   A. Diego is at trophic levels 2 and 4; Darius is at trophic levels 2, 3, and 4.

   B. Diego is at trophic levels 3 and 4; Darius is at trophic levels 2 and 3.

   C. Diego is at trophic levels 1 and 3; Darius is at trophic levels 2 and 3.

   D. Diego is at tropic levels 2 and 3; Darius is at trophic levels 1 and 2.

47. Which of the following pairs of organisms can form a relationship that allows the earliest colonization of a barren environment, such as a new volcanic island?

   A. An insect and a moss

   B. A fungus and an algae

   C. A moss and a fungus

   D. An insect and a protozoan

Questions 48-51 refer to the following cladogram:

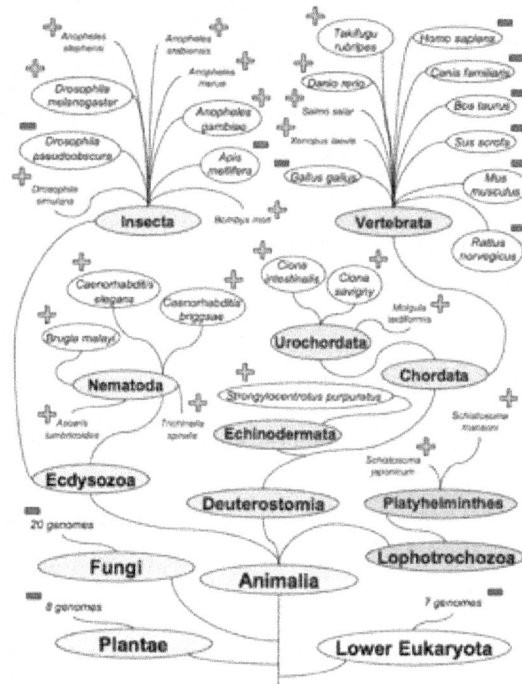

48. According to this "tree of life" view of evolution, animals are divided into three major groups, or superphyla. Which of these three groups includes humans?

    A. Vertebrata

    B. Chordata

    C. Lophotrochozoa

    D. Deuterostomia

49. The term "Deuterostomia" means "mouth second." Why is this name used for this group?

    A. During the embryonic development of the Deuterostomia, the second gastrulation of the hollow zygote takes place at the mouth.

    B. During the embryonic development of the Deuterostomia, gastrulation of the hollow blastula begins at the anus and ends at the mouth.

    C. During the embryonic development of the Deuterostomia, gastrulation of the hollow blastula begins at the feet and ends at the mouth.

    D. During the embryonic development of the Deuterostomia, gastrulation of the hollow gastrula begins at the anus and ends at the mouth.

50. After this cladogram was published, one of the three superphyla was divided into two separate superphyla. The new superphylum is the Platyzoa, and includes flatworms. Of the animals in the cladogram above, which genus would be moved to this new superphylum?

   A. *Caenorrhabditis*

   B. *Trichinella*

   C. *Schistosoma*

   D. *Ascaris*

51. Which of the following groups contains spiders?

   A. The Ecdysozoa

   B. The Lophotrochoza

   C. The Insecta

   D. The Echinodermata

52. A certain protein is destined to be transported to another part of the body. Which of the following will be involved in packaging the protein after gene expression?

   A. The Golgi apparatus

   B. The rough endoplasmic reticulum

   C. The ribosome

   D. Protein packagase, a transmembrane protein

53. Cytochrome c oxidase is a large, transmembrane protein that is an integral part of the electron transport chain. Which of the following describes cytochrome c oxidase?

   A. Hydrophobic polypeptide with nonpolar amino acids to interact with water, with metal atoms

   B. Hydrophilic polypeptide with polar amino acids to interact with water, with cytochrome atoms

   C. Alternating hydrophobic and hydrophilic polypeptide regions, with helium atoms

   D. Alternating hydrophobic and hydrophilic polypeptide regions, with metal atoms

54. Noncoding RNA is RNA that will not be used as a template for building a polypeptide. Which of the following are examples of noncoding RNA?

A. ribosomal RNA, messenger RNA, and small nuclear RNA

B. messenger RNA, tRNA, and ribosomal RNA

C. micro RNA, tRNA, ribozymes

D. small nuclear RNA, cDNA, and ribozymes

55. Euglena are single-celled organisms that show both plant- and animal-like properties. They can photosynthesize, but also eat other organisms and can move through their freshwater environment using whip-like cellular appendages. Knowing this, which of the following cellular structures do Euglena have?

A. Chloroplasts and a cell wall

B. Flagella and chloroplasts

C. Flagella and gap junctions

D. Chloroplasts and plasmodesmata

56. Some neurons are very long, for example, neurons that extend from the spine to muscles in your feet. To enable a signal to travel this length quickly enough to allow efficient movement, the axons are coated with insulating cells that allow the signal moving down the axon to move in a jumping/sliding motion called saltatory conduction. What are these insulating cells called, what substance surrounds the axon, and what happens in the gap in between them?

A. adipocytes, myelin, membrane potential

B. neurons, glia, action potential

C. neuroglia, myelin, action potential

D. microglia, myelin, membrane potential

57. In order to colonize land, plants had to develop separate systems for obtaining sunlight and obtaining water and nutrients. Seaweeds, in contrast, live in water, which allows the whole plant surface to have access to both sunlight and water/minerals. Knowing this, what is the likely evolutionary history of sea grass, flowering plants that live completely underwater in shallow marine environments like coral reefs?

A. Nonvascular land plants evolved from algae. Then, some of these eventually evolved into flowering land plants. Some of these then evolved to be able to survive underwater in shallow marine habitats.

B. Nonvascular marine plants evolved from algae. Then, some of these eventually evolved into vascular marine plants. Some of these then evolved into flowering marine plants.

C. Multicellular algae (seaweed) evolved from unicellular algae. Then, some multicellular seaweeds evolved into flowering seaweeds. Next, some of these gained a grassy morphology optimized for shallow marine habitats.

D. Multicellular algae (seaweed) evolved from unicellular algae. Then, some of these evolved into nonvascular marine plants. Then, some of these evolved into flowering plants with a grassy morphology optimized for shallow marine habitats.

58. During the Calvin cycle of photosynthesis, plants create glucose using the energy in sunlight to build glucose molecules. What is the net reaction of this process?

A. $6\ CO_2 + 6\ O_2 => \text{light energy} => C_6H_{12}O_6 + 6\ H_2O$

B. $6\ CO_2 + 6\ H_2O => \text{light energy} => C_6H_{12}O_6 + 6\ O_2$

C. $C_6H_{12}O_6 + 6\ O_2 => \text{light energy} => 6\ CO_2 + 6\ H_2O$

D. $C_6H_{12}O_6 + 6\ O_2 => 6\ CO_2 + 6\ H_2O + \text{energy}$

59. Hookworms are roundworms that hook into the intestinal lining and suck blood for their nourishment. What type of organism are hookworms?

A. Protozoan parasite

B. Protozoan symbiont

C. Metazoan parasite

D. Metazoan mutualist

60. The protozoan parasitic amoeba *Entamoeba histolytica* causes a serious disease, but molecular studies have found that what was thought to be a single organism is in fact two: the pathogenic *Entamoeba histolytica* and a non-pathogenic commensal, *Entamoeba dispar*. Which animal is analogous in lifestyle to *Entamoeba dispar*?

    A. *Plasmodium falciparum*, the protozoan that causes malaria

    B. *Demodex folliculorum*, a mite that lives on the skin of almost all humans and eats oil from sebaceous glands

    C. Intestinal bacteria, which play a role in synthesizing vitamin B and vitamin K as well as metabolizing bile acids, sterols and xenobiotics

    D. *Paragonimus westermani*, a lung fluke that causes the human disease paragonimiasis

61. **Which of the following describe communication between two neurons?**

    A. A neurotransmitter is released into the synapse, it binds a receptor on the postsynaptic neuron, and this generates an action potential.

    B. A neurotransmitter is released into the synapse, it binds to a receptor on the postsynaptic neuron, and this generates a membrane potential.

    C. A hormone is released into the synapse, it binds to a receptor on the postsynaptic neuron, and generates an action potential.

    D. A cytokine is released into a gap junction, where it travels to the postsynaptic neuron, and generates a membrane potential.

62. **Male mosquitoes have large, feathery antennae and mouthparts adapted to a diet of nectar, whereas female mosquitoes have much smaller antennae and long, syringe-like mouthparts. Why?**

    A. Males can survive on flower nectar, but females require the extra protein of a blood meal in order to lay eggs. Males with bigger antennae are better at detecting females.

    B. Females can survive on flower nectar, but they require the extra protein of a blood meal in order to lay eggs. Females with bigger antennae are better at attracting males.

    C. There is no adaptive advantage to either trait. The evolution of these traits is the result of genetic drift.

    D. The long, sucking mouthparts of the females enable them to obtain more calories than afforded by flower nectar alone. The large antennae of males is the result of higher levels of testosterone.

63. **Which molecule allows mosquitoes to detect humans or other mammals that can provide a blood meal?**

    A. O_2

    B. ATP

    C. CO_2

    D. $C_6H_{12}O_6$

Part B. Grid-in Questions

1. A research lab is using mutant mice for its experiments to determine the effects of a certain gene deficiency on blood pressure. Each month, 200 mice are born and 45 mice die. The mouse colony currently has 630 mice. What is the growth rate of the colony each month? Round your answer to the hundredth.

2. A green-fluorescent protein exists that will localize to the mitochondrial membrane. A researcher would like to visualize the mitochondria in murine macrophages. She has a stock solution of the stain at a concentration of 2μM and the protocol calls for a final concentration of 50nM in a volume of 200 microliters. How many microliters of stock solution does the researcher need?

3. Evaporation of water from leaves of plants to the environment is called transpiration. The graph below shows the transpiration rates of an avocado tree throughout the year. How many liters of water does an avocado tree lose during the summer (June through August)?

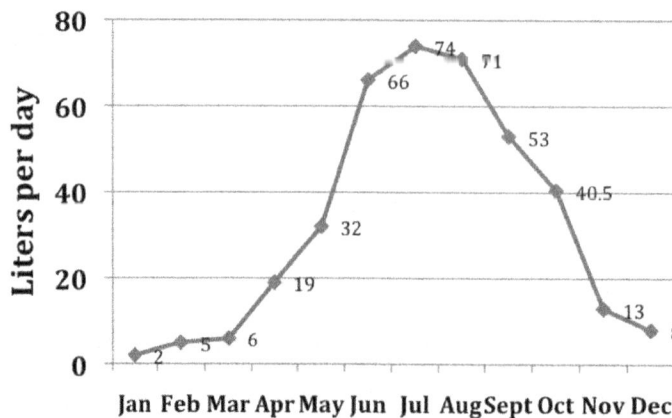

4. In a particular species of sloth, there is a recessive allele of a gene that results in an unusually long tail. This allele occurs at a frequency of 0.527 in the population. According to the Hardy-Weinberg equation, what is the frequency of long-tailed sloths? Round your number to the nearest thousandth.

5. During vasculogenesis (the development of blood vessels), cells called angioblasts are differentiated into endothelial cells. These endothelial cells migrate in response to growth factors to their correct location and will become the inner lining of capillaries. If a single angioblast differentiates into an endothelial cell, and the subsequent rate of division is once every 4 minutes, how long will it take to produce enough cells to line a capillary requiring 262, 144 cells? Report your answer in hours.

6. Erythropoietin is a hormone that controls red blood cell production. Translation of the erythropoietin gene results in a protein containing 193 amino acids. This amino acid chain is later cleaved to produce a chain that is 166 amino acids. How many ATP molecules are required for the translation of erythropoietin?

Free Response (6 short, 2 long) _____

Short Free Response:

1. Imagine that you are digesting a grilled cheese sandwich. Name and briefly describe the function of three enzymes that will be used to break down the sandwich, and name and briefly describe the function of three organs and their physiological function involved in the digestive process. (Consider only digestion and not absorption of nutrients or excretion of waste.)

2. Describe the role of oxygen in cellular respiration. What will occur if oxygen is not available?

3. Explain the differences between microtubules and microfilaments.

4. A deficiency of proteins responsible for nuclear membrane reassembly prevents cells from properly undergoing mitosis. List the stages of mitosis and explain what will happen in the case of such protein deficiency.

5. Explain the differences between meiosis and mitosis.

6. Describe the physiological aspects of glucose processing during cellular respiration, with an emphasis on the locations in which the process occurs.

Long Free Response:

7. Photosynthesis is crucial to the survival of plants, algae, and cyanobacteria.

    (A) Outline the light-dependent reactions of photosynthesis.

    (B) Explain how temperature could affect these processes.

    (C) Graph the expected results.

8. In order to understand the processes of cellular respiration in muscle cells grown in a laboratory cell culture, the cells were lysed and fractionated to isolate the mitochondria from the cytoplasm. The fractionated samples were treated with glucose or pyruvate for 1 hour, and at the end of the incubation period, each sample was analyzed for the presence of lactic acid or carbon dioxide. The questions below should be answered based on the results presented here:

Cell fraction	CO_2	Lactic Acid
Mitochondria incubated with glucose	Absent	Absent
Mitochondria incubated with pyruvate	Present	Absent
Cytoplasm incubated with glucose	Absent	Present
Cytoplasm incubated with pyruvate	Absent	Present

a. What does the presence of lactic acid in the samples suggest about which process is occurring in each particular cell fraction?

b. Why was carbon dioxide not produced by the cytoplasmic fraction incubated with glucose?

c. Why did the cytoplasmic fraction produce lactic acid in the presence of both glucose and pyruvate?

d. Why did the mitochondrial fraction produce carbon dioxide in the presence of pyruvate but not in the presence of glucose?

e. Draw the chemical reaction occuring in each scenario.

Question Number	Correct Answer	Your Answer	Question Number	Correct Answer	Your Answer
1	A		33	A	
2	A		34	B	
3	C		35	A	
4	C		36	D	
5	A		37	B	
6	D		38	D	
7	B		39	A	
8	B		40	C	
9	B		41	A	
10	C		42	B	
11	C		43	D	
12	A		44	C	
13	D		45	B	
14	C		46	A	
15	C		47	B	
16	B		48	D	
17	A		49	B	
18	C		50	C	
19	B		51	A	
20	B		52	A	
21	C		53	D	
22	D		54	C	
23	A		55	B	
24	A		56	C	
25	C		57	A	
26	D		58	B	
27	B		59	C	
28	C		60	B	
29	D		61	B	
30	C		62	A	
31	B		63	C	
32	D				

Practice Test Two

1. **During pregnancy, the body undergoes a number of changes to facilitate the birth of the baby. Among these are changes that allow the bones of the pelvis to be pulled apart to widen the birth canal. Which of the following structures would be expected to relax to accomplish this?**

    A. Ligaments

    B. Tendons

    C. Muscles

    D. Bones

    Answer: A.

    Ligaments. Ligaments connect bone to bone. If they relax, the bones can pull apart more.

2. ***Dalbulus maidis* is an insect pest of domesticated maize (corn) and its wild ancestor, teosinte, both of which originated in central Mexico. *D. maidis* eggs are laid on the midrib and leaf blade in teosinte but only on the midrib of the upper leaf surface in maize. A Mexican biologist observed that larvae of another pest, the chewing insect *Spodoptera frugiperda*, consume the leaf blade of teosintes and maize, but not the midrib, and he thought that this might explain the egg-laying behavior of *D. maidis*. Which of the following could explain his conclusion?**

    A. Maize, planted in dense fields, hosts larger populations of pests. In the context of a large population of *S. frugiperda*, the *D. maidis* eggs are at risk of being eaten by *S. frugiperda* if they are laid on the blade. This selection pressure has led to niche partition between the two species.

    B. Teosinte, in its diverse natural habitat, hosts a smaller population of pests. This provides plenty of food for all of the insects, and this selection pressure has led to niche specialization by the two species.

    C. As it has evolved into a different species from its ancestor teosinte, maize has evolved chemical defenses to repel insect pests. The selection pressure of the different insects has led to specialized defenses that are toxic to *D. maidis* on the midrib and *S. frugiperda* on the leaf blade.

    D. Maize has a much more diversified herbivore population, including *D. maidis* and *S. frugiperda*, since it is more genetically heterogeneous and evolutionarily basal to teosinte. The dependence of maize on insect herbivores has given rise to selection pressure that has led to niche partition between the two species.

    Answer: A.

    Maize, planted in dense fields, hosts larger populations of pests. In the context of a large population of *S. frugiperda*, the *D. maidis* eggs are at risk of being eaten by *S. frugiperda* if they are laid on the blade. This selection pressure has led to niche partition between the two species. Cultivated crops lead to high populations of specialized pest insects. In the context of a large number of chewing caterpillars, eggs laid on the leaf blades are at risk of being eaten. Since the larvae do not eat the leaf midrib, this is a safe place to lay eggs. Evolutionary pressure will select this behavior. In the wild, there are not enough *S. frugiperda* larvae to pose a great enough danger to exert selection pressure. B and D are essentially nonsense. C sounds plausible, except that if there was an *S. frugiperda*-specific toxin expressed in the leaf blade and a *D. maidis*-specific toxin on the midrib, this would lead to the opposite situation to what was observed—*S. frugiperda* would eat the midribs and *D. maidis* would lay eggs on the leaf blades, not the other way around!

3. **Urinary tract infections start when bacteria outside the body enter the urinary system and move up into the body. Choose the correct order of structures in women that might be invaded and lead to infection.**

   A. The urethra, then the bladder, then the ureter, eventually leading to a liver infection

   B. The vagina, then the bladder, then the urethra, eventually leading to a kidney infection

   C. The urethra, then the bladder, then the ureter, eventually leading to a kidney infection

   D. The vagina, then the urethra, then the ureter, eventually leading to a bladder infection

   Answer: C.

   The urethra, then the bladder, then the ureter, eventually leading to a kidney infection. The urethra is the opening where the bacteria enter, then they travel to the bladder, then they could travel up the ureter, which connects the bladder with the kidney, and from there eventually infect the kidney. This is the same for women and men. This answer could be deduced by eliminating the other three, since none of those structures are connected in the given order.

For questions 4-7, refer to the following phylogram. This phylogenetic tree is based on the reverse transcriptase amino acid sequence of several LTR retrotransposons, viral-like mobile genetic elements. These types of retrotransposons are similar to retroviruses, except that they lack the viral envelope needed to be transmitted from one host organism to another (to be infectious). They are also known as endogenous retroviruses.

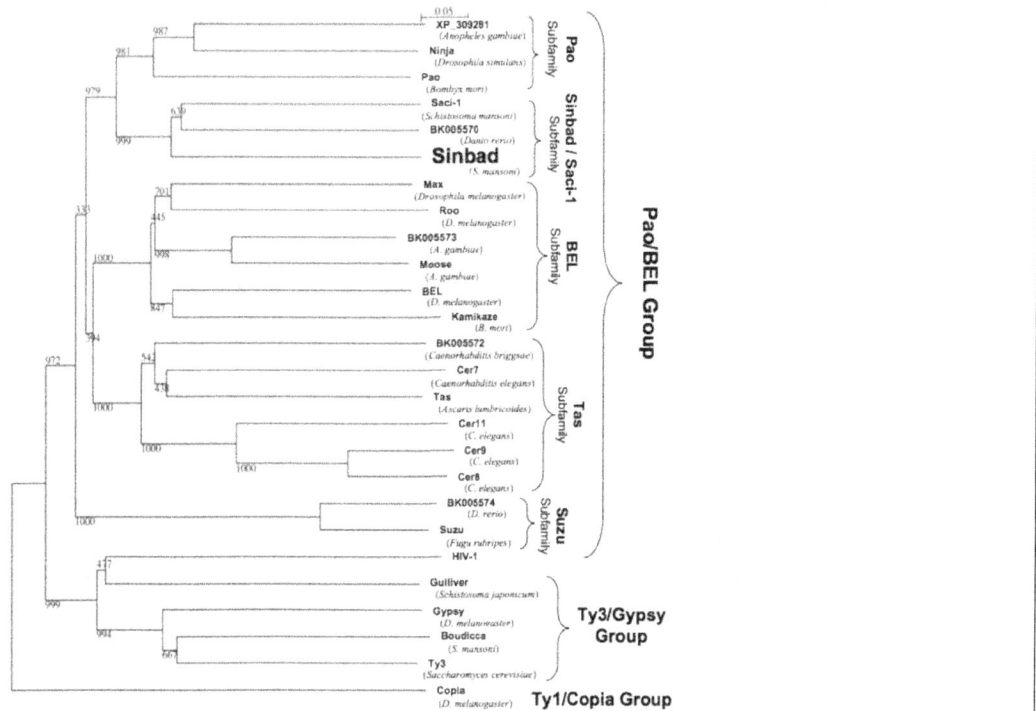

4. In the phylogram above, which of the following is the most evolutionarily distant from the Ty3 element?

A   Pao

B. HIV-1

C. Copia

D. Sinbad

Answer: C.

Copia. The Ty1/Copia group is basal to the tree; its position outside the branching of the other elements shows this.

5. **Which of the following is most closely related to the *Boudicca retrotransposon*, a retrotransposon from the African blood fluke *Schistosoma mansoni*?**

A. HIV-1, from *Homo sapiens*

B. Copia, from *Drosophila melanogaster*

C. Cer7, from *Caenorhabditis elegans*

D. Sinbad, from *Schistosoma mansoni*

Answer: A.

HIV-1, from *Homo sapiens*. HIV-1 is also within the Ty3/Gypsy group of retrotransposons. The others are all from more evolutionarily distant groups.

6. **The closest relative of the Sinbad retrotransposon from *Schistosoma mansoni*, a parasitic flatworm, is very close on the phylogram to a sequence from *Danio rerio*, the zebrafish. Which of the following is one possible explanation for this?**

A. *Schistosoma mansoni* is very closely related to *Danio rerio*.

B. The flatworms, phylum Platyhelminthes, evolved from fish.

C. *Schistosoma mansoni* and *Danio rerio* both evolved Sinbad-like retrotransposons independently; this is known as convergent evolution.

D. A Sinbad-like retrotransposon may have been horizontally transmitted from an ancestor of *S. mansoni* to an ancestor of *Danio rerio*.

Answer: D.

A Sinbad-like retrotransposon may have been horizontally transmitted from an ancestor of *S. mansoni* to an ancestor of *Danio rerio*. This could have occurred because a parasitic worm could be in close enough contact with its host for a viral-like element to enter a host cell and integrate into the genome. C is unlikely, since convergent evolution occurs in response to selection pressure, and results in a similar phenotype but different genotype. A and B are simply not true and do not make sense considering the general evolution of animals.

7. *Anopheles gambiae* is a mosquito, *Drosophila* spp. are fruit flies, *Bombyx mori* is the silkworm (a moth), *Schistosoma* spp. are parasitic flatworms, *Danio rerio* and *Fugu rubripes* are fish, *Ascaris* is a parasitic roundworm, and *Caenorhabditis* spp. are free-living roundworms. Which of the Pao/BEL group subfamilies include retrotransposons that are most likely to have undergone horizontal transmission?

A. Pao

B. Sinbad

C. Tas

D. Suzu

Answer: B.

Sinbad. Only the Sinbad subfamily has radically different host genomes— flatworms and fish. Pao elements are all in insects, Tas elements are all in roundworms, and Suzu elements are all in fish. Sinbad-like elements, however, are in flatworms and fish, making it unlikely that the elements were present in a common ancestor.

8. **If a person has gallbladder cancer, on the basis of the pain that they feel, what organ might a doctor mistakenly think is diseased?**

A. The kidneys

B. The liver

C. The appendix

D. The heart

Answer: B.

The liver. The liver is right next to the gallbladder. Gallbladder pain would be in the upper right hand side of the belly, where the liver is. The kidneys are towards the back; the appendix is in the lower belly and the heart is in the upper chest.

9. **The wild relatives of maize (corn) have trichomes on their early leaves, whereas commercial maize has lost this trait. Recent research from the University of Guadalajara suggests that this early pubescence my serve as a physical barrier to thwart insect pest herbivores and protect the young plants. What does the word "pubescence" mean in this context?**

   A. Adolescence

   B. Being hairy

   C. Being mature enough to produce seeds

   D. Producing a natural insecticide

   Answer: B.

   Being hairy. Pubescence means having downy or short, fine hair. It has been used to mean being in the state of adolescence in humans, for obvious etymological reasons, but in this botanical context it has its straightforward, original meaning. You could also deduce this answer by either knowing that trichomes are plant hairs, or by substituting the other answers into the sentence; biologically, none of the others make sense.

10. **Which type of snack would require the most amylase activity for digestion?**

    A. Hard candy

    B. Dried meat (jerky)

    C. Toast

    D. Butter

    Answer: C.

    Toast. Amylase breaks starch into sugars. Bread is high in starch. Hard candy contains sugars (amylase does not break down sucrose), dried meat contains mainly protein, and butter consists mainly of fat.

11. **In vascular plants, the dominant part of the plant is diploid. Which of the following statements about vascular plants are true?**

    A. The sporophytes of vascular plants create gametes by meiosis.

    B. The sporophytes of vascular plants create gametes by mitosis.

    C. The sporophytes of vascular plants create spores by meiosis.

    D. The sporophytes of vascular plants create spores by mitosis.

    Answer: C.

    The sporophytes of vascular plants create spores by meiosis. The sporophytes create spores by meiosis, and the resulting haploid spores grow into haploid gametophytes, the nondominant stage. The gametophytes then produce gametes by mitosis.

12. **Human kids are well-known to hate vegetables and spicy food like chili peppers, but adult humans often love these flavors. How can this be explained, in terms of physiology and evolution?**

    A. Plants have evolved compounds during phylogeny to repel herbivores but the human body learns during ontogeny that the compounds are healthy, leading to a preference for the flavors.

    B. Plants have evolved compounds during ontogeny to repel herbivores but the human body learns during phylogeny that the compounds are healthy, leading to a preference for the flavors.

    C. Plants have evolved compounds during phylogeny that enhance human health, but these compounds are not healthy for children.

    D. Children have evolved a repulsion for bitter compounds during phylogeny but plants have evolved compounds that appeal to adults during ontogeny.

    Answer: A.

    Plants have evolved compounds during phylogeny to repel herbivores but the human body learns during ontogeny that the compounds are healthy, leading to a preference for the flavors. Phylogeny is the evolution of traits over generations through natural selection; ontogeny is the development of traits (phenotypes) over the course of an individual's lifetime. Plants (vegetables) evolve bitter compounds to prevent themselves from being eaten by animals like humans. The vegetables are very healthy for humans, so the human body will develop a preference for these flavors since they are associated with greater health (as long as the human actually eats the vegetables--even if by coercion from the parents!).

13. **Kidney stones form through the precipitation of dissolved calcium and oxalate in the kidney under acidic conditions. A high animal protein intake significantly increases excretion of calcium, oxalate, and uric acid into the urine. Which of the following will provide the BEST reduction of kidney stones?**

    A. A low-meat diet and a high fluid intake

    B. A vegetarian diet and a low fluid intake

    C. A high-meat diet and a high fluid intake

    D. A vegan (no meat or dairy products) diet and a high fluid intake

    Answer: D.

    A vegan (no meat or dairy products) diet and a high fluid intake. Vegans have been shown to have the lowest rates of kidney stones, presumably because their low intake of animal protein decreases the excretion of chemicals that can precipitate into kidney stones. A high fluid intake will also decrease kidney stone risk, because when the chemicals are dissolved in a greater volume of fluid, they will be less likely to precipitate.

14. **Researchers at the University of Guadalajara found that two species of teosinte, the wild ancestor of corn, expressed trichomes at an early developmental stage whereas modern commercial maize (corn) did not express this phenotype until a later developmental stage. The trichomes, fine hairs growing out of the leaf blades, prevented chewing by the generalist pest insect *Spodoptera frugiperda*, resulting in significantly less damage in the teosintes than in the modern maize. Which of the following is most likely to be true?**

A. The two wild species independently developed the early-trichome trait, since both were under selection pressure from the herbivore, *Spodoptera frugiperda*.

B. One wild species developed the early-trichome trait under pressure from the herbivore, and passed it to the other wild species by conjugation.

C. The wild species from which modern maize developed had the early-trichome trait but modern maize lost it because pesticide use reduced selection pressure for the trait.

D. Modern maize had the early trichome trait, then passed it to one of the wild species through a transposon.

Answer: C.

The wild species from which modern maize developed had the early-trichome trait but modern maize lost it because pesticide use reduced selection pressure for the trait. Selection pressure will favor the development of a defense, like trichomes. However, if the selection pressure is lifted, such as when humans use pesticides to kill the insects, the organism can evolve to lose the trait. (If a mutation happens that causes loss of the trait-- which can easily occur-- that mutation will remain in domesticated cornfields where there is pesticide use.)

15. **Considering the anatomy of the human body, if a kidney stone is large enough to fill the renal pelvis, which of the following is true?**

A. The stone can be passed by giving the patient a large amount of fluid.

B. The stone is too large to be passed; it will cause serious pain and eventually gallbladder dysfunction.

C. The stone must be disrupted (e.g. by sound waves) or removed surgically.

D. An alpha blocker should be administered to relax the muscles of the ureter and then the stone can be passed.

Answer: C.

The stone must be disrupted (e.g. by sound waves) or removed surgically. For this question, you need to understand that the inside of the renal pelvis is much larger than the lumen of the ureter. If a stone is large enough to fill the renal pelvis, it will have to be broken up or removed surgically; it will be too large to pass, even with alpha blockers. (The gallbladder is related to digestion, not urine, so B is not correct.)

For questions 16-20 refer to the drawing and description below:

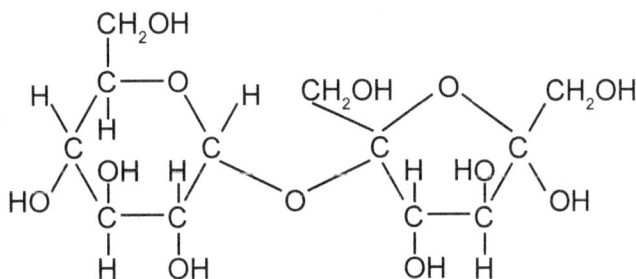

Under anaerobic conditions, yeast will ferment this compound to produce energy. However, before this can happen, they must break this disaccharide into two monosaccharides, glucose and fructose, both of which have the same molecular formula: $C_6H_{12}O_6$. The unbalanced equation of glucose or fructose fermentation is:

$$C_6H_{12}O_6 \rightarrow C_2H_5OH + CO_2$$

16. In order to break this disaccharide into two monosaccharides, what type of small molecule will participate in the reaction, adding its atoms to the final two monosaccharides?

A. CO_2

B. H_2O

C. $C_6H_{12}O_6$

D. CH_3COOH

Answer: B.

H_2O. During hydrolysis, H_2O reacts with the disaccharide, resulting in -OH groups on each monosaccharide in place of the O between them.

17. What enzyme might catalyze the metabolism of sucrose into glucose and fructose?

A. Glycoside hydrolase

B. Protease

C. Lipase

D. Chantililase

Answer: A.

Glycoside hydrolase. You could either know this, or realize that protease breaks proteins, lipase breaks fats, and chantililase is nonsense. Two other clues: 1. The prefix "gly" refers to sweet things, either sugars, sweet amino acids, or something related to them. (For example, the breakdown of glucose is glycolysis.) 2. Hydrolase is an enzyme that catalyzes hydrolysis by breaking apart water. Since water must be broken apart and incorporated into the new fructose and glucose molecules, this also is a strong clue.

18. **When a winemaker makes wine using yeast fermentation, what are the products of the fermentation of a disaccharide like that pictured above?**

    A. Fructose and glucose

    B. Ethanol and water

    C. Ethanol and carbon dioxide

    D. Fructose and ethanol

    Answer: C.

    Ethanol and carbon dioxide. This answer can be arrived at in a number of ways, but it is obvious from the equation above, that carbon dioxide is one product. Since C is the only choice that includes carbon dioxide, even if you do not know anything else about fermentation, this is clearly the correct choice.

19. **How would the energy yield of anaerobic fermentation compare with the energy yield from the same disaccharide under aerobic conditions?**

    A. Fermentation yields much more energy.

    B. Fermentation yields much less energy.

    C. Fermentation yields the same amount of energy.

    D. Fermentation yields a little less energy.

    Answer: B.

    Fermentation yields much less energy. Fermentation yields only 2 ATP molecules from each glucose, whereas aerobic metabolism yields about 30 molecules of ATP (theoretical yield: 36 molecules of ATP)

20. **If the yeast were given the same disaccharide under aerobic conditions, which of the following would be involved in the metabolism that would take place under these conditions?**

    A. The Calvin cycle, mitochondria, electron transport chain

    B. The Krebs cycle, mitochondria, chemiosmotic potential

    C. The Krebs cycle, electron transport chain, carbon fixation

    D. The Krebs cycle, electron transport chain, fermentation,

    Answer: B.

    The Krebs cycle, mitochondria, chemiosmotic potential. The Krebs cycle is the main cycle for aerobic respiration, it takes place in the mitochondria, and most of the energy is harvested by ATP synthase converting the chemiosmotic potential of H^+ ions piled on one side of a membrane (accomplished by the electron transport chain) into bond energy in ATP.

21. **Can human bodies harvest energy from the ethanol in wine?**

    A. No, ethanol cannot be fermented because it is already the product of fermentation.

    B. No, ethanol can only be detoxified using alcohol dehydrogenase.

    C. Yes, ethanol can be processed and then enter the Krebs cycle and be metabolized to CO_2.

    D. Yes, ethanol can be fermented via anaerobic respiration.

    Answer: C.

    Yes, ethanol can be processed and then enter the Krebs cycle and be metabolized to CO_2. Ethanol is processed into acetic acid, which can then enter the Krebs cycle and be metabolized to CO_2.

22. **Two closely related crop plants produce a natural flavor compound that is appealing to humans and also detectable by a common insect pest. One subspecies (Plant A) begins producing the flavor compound at the 2-leaf stage of development. The other subspecies (Plant B) does not begin producing it until the 8-leaf stage, but produces much more of it than Plant A. Plant B requires substantially more chemical pesticide than Plant A to reduce damage from this insect to produce a good crop yield. What does this say about the feeding preferences of the insect?**

    A. The compound is toxic to the insect, which prefers to eat older plants.

    B. The insect likes the flavor of the compound, and prefers to eat younger plants.

    C. The insect is repelled by the compound, and prefers to eat older plants.

    D. The insect is repelled by the compound, and prefers to eat younger plants.

    Answer: D.

    The insect is repelled by the compound, and prefers to eat younger plants. Plant A requires less pesticide because the compound is repelling the insect. Since Plant B produces more of the compound than Plant A when it is older, but Plant A produces more of the compound than Plant B when it is younger, the insect must prefer younger plants. (The young shoots/leaves of Plant A have the bad-tasting compound, whereas the young shoots/leaves of Plant B taste good until it gets older.)

23. **Textbooks often say that respiration yields 38 molecules of ATP, but the real yield is closer to 30 molecules of ATP. What might account for this?**

A. Leaky membranes

B. Non-metabolic mitochondria

C. Partial anaerobic metabolism/fermentation

D. Some of the ATP is used to power the Krebs cycle

Answer: A.

Leaky membranes. Most of the ATP comes from a gradient of H^+ ions across a membrane; as they flow towards the side with a low concentration of H^+, they must go through ATP synthase, which harvests this potential energy and converts it to bond energy in ATP (think of a waterwheel for an old mill or modern hydroelectric dam). If the membrane is leaky, it's like a leaky dam—the H^+ ions will flow through the leaky membrane and not produce ATP in the process.

Questions 24-27. Parasitoid wasps can be useful in natural pest control, since many are natural predators of crop pest insects. A team of researchers from the USDA and the Centro de Ecologia in Mexico, looking for an all-female strain of parasitoids, found a strain of the fruit fly parasitoid *Odontosema anastrephae* that included only females, with no males after several generations. The following figure is a phylogenetic tree of the cytochrome C oxidase subunit 1 (COI) gene of individuals of the parthenogenetic (all-female) and arrhenotokous (male and female) strains. The numbers on the branch node are "bootstrap values"—out of 1,000 repetitions of the tree-drawing process, this is how many times the computer has given that particular tree configuration.

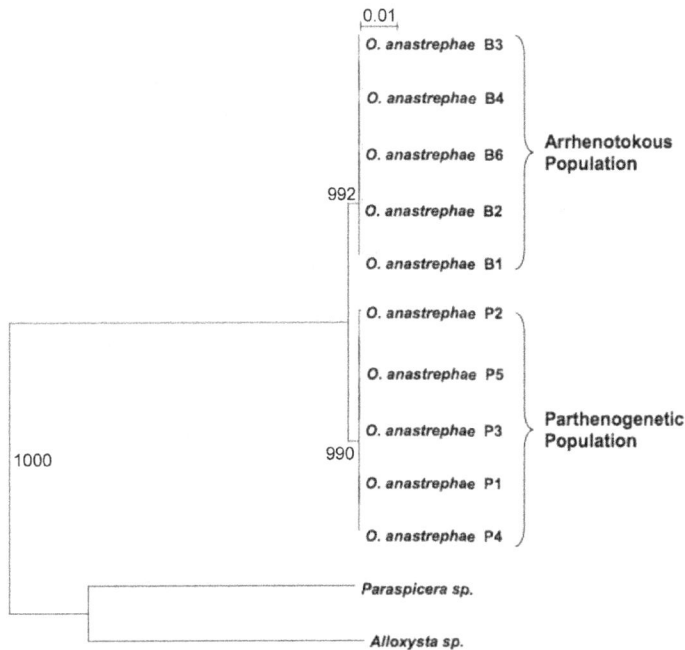

24. **Based on the phylogram above, what could you conclude about these wasps?**

   A. The wasps represent two populations that are closely related but genetically distinct; they probably will not interbreed if given the opportunity.

   B. There is no significant difference between the two populations; they will probably interbreed if they are given the opportunity.

   C. The two populations are genetically distinct; they actually are different species. They will definitely interbreed if given the opportunity.

   D. The parthenogenetic wasps are a subspecies of the arrhenotokous wasps, but this branch has recently diverged so they will probably interbreed if given the opportunity.

Answer: A.

The wasps represent two populations that are closely related but genetically distinct; they probably will not interbreed if given the opportunity. The two groups are different, since the individuals cluster together with bootstrap values of over 900 (Eliminate answer B). However, the branch lengths are very short, indicating that the two groups are closely related (eliminate answer C). Answer D, which population diverged from which, cannot be determined on the basis of this phylogram.

25. **Considering the reproductive strategy of these wasps, what must be true of the parthenogenetic *O. anastrephae*?**

A. In the parthenogenetic strain, mating-type alpha female wasps reproduce sexually with mating-type alpha females.

B. Female wasps of the parthenogenetic strain only have one parent, but will nevertheless be unique because of the effect of crossing over.

C. Wasps of the parthenogenetic strain are clones – either full or half-clones.

D. Male wasps of parthenogenetic strain are clones of their mother.

Answer: C.

Wasps of the parthenogenetic strain are clones – either full or half-clones. Since these wasps only have one parent, they are either full clones of the parent (genetically identical) or, if after meiosis the genetic material in the haploid cell is simply doubled, they will be half-clones of their mother.

26. **USDA biologists were looking for all-female wasps because it would cut production costs of these natural predators in half. Knowing this, what is probably true about these predators?**

A. Both male and female wasps eat many insect larvae per day, but females eat more than males.

B. Male wasps are more expensive to produce because they are larger.

C. The wasps parasitize the pest insects by burrowing into the larvae.

D. The wasps parasitize the pest insects by laying eggs inside the larvae.

Answer: D.

The wasps parasitize the pest insects by laying eggs inside the larvae. If production costs will be cut in half by eliminating males, the males must not be functioning as predators at all. The only choice that is a function only females can accomplish is D.

27. **The cytochrome C oxidase subunit I gene is often used in phylogenetic comparisons like the one in the previous questions. Why?**

    A. Cytochrome C functions as part of the immune system, and is therefore highly conserved in all eukaryotes.

    B. Cytochrome C functions as part of the electron transport chain, and is therefore highly conserved in all eukaryotes.

    C. Cytochrome C functions as part of the Calvin cycle, and is therefore highly conserved in all eukaryotes.

    D. Cytochrome C is only found in insects and is therefore useful for distinguishing between subspecies.

    Answer: B.

    Cytochrome C functions as part of the electron transport chain, and is therefore highly conserved in all eukaryotes. Since cytochrome C is part of the electron transport chain, all eukaryotes have this protein. Mutations will tend to be lethal because a well-functioning electron transport chain is critical for survival, so relatively few mutations are passed on and the cytochrome C genes are similar enough to be comparable between even in evolutionarily distant species.

28. **Which of the following statements are true?**

    A. Fungi are to slime molds as plants are to algae.

    B. Plants are to algae as cyanobacteria are to mitochondria.

    C. Cyanobacteria are to chloroplasts as eubacteria are to mitochondria.

    D. Archaea are to mitochondria as cyanobacteria are to chloroplasts.

    Answer: C.

    Cyanobacteria are to chloroplasts as eubacteria are to mitochondria. According to endosymbiotic theory, chloroplasts evolved from cyanobacteria and mitochondria evolved from eubacteria.

29. **Plasmodium is a protozoan parasite that causes malaria. Which of the following is true about Plasmodium?**

A. It is a multicellular member of the kingdom Animalia in the older system of classification, and domain Archaea in the modern three-domain system of classification.

B. It is a multicellular member of the kingdom Protista in the older system of classification, and domain Eubacteria in the modern three-domain system of classification.

C. It is a unicellular member of the kingdom Monera in the older system of classification, and domain Eubacteria in the modern three-domain system of classification.

D. It is a unicellular member of the kingdom Protista in the older system of classification, and domain Eukarya in the modern three-domain system of classification.

Answer: D.

It is a unicellular member of the kingdom Protista in the older system of classification, and domain Eukarya in the modern three-domain system of classification. Protozoans are unicellular eukaryotic organisms.

30-33. Luciferase is an enzyme that produces light in fireflies. The amount of luciferase expressed in a cell can be deduced by measuring its light output in a luminometer. A biologist was interested in knowing whether the promoters for two retrotransposons she had discovered, Boudicca and Sinbad, were functional. Molecular biologists make use of plasmids, small circular DNAs that are naturally exchanged between bacteria, to look at gene expression. Stretches of DNA the researcher is interested in studying are inserted into the plasmid, and the biologists use lab techniques to get bacteria to take up the plasmids. Questions 30-33 refer to the following figure:

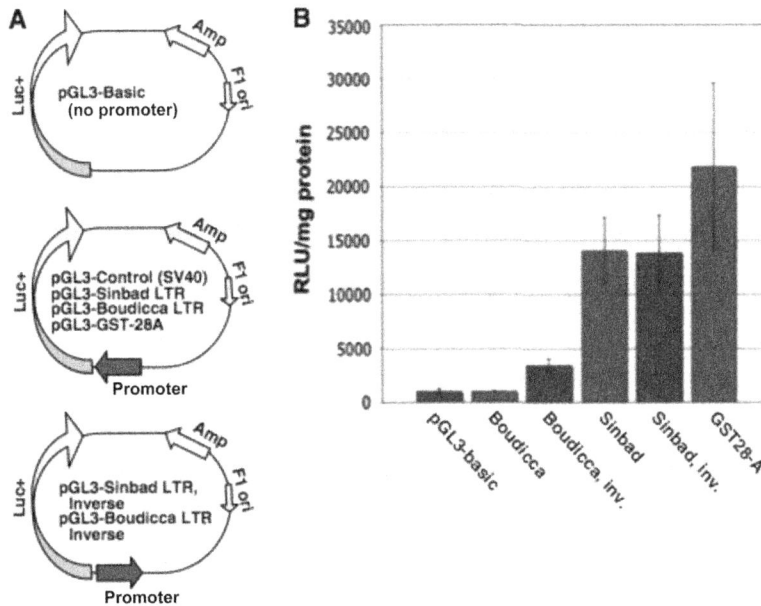

30. Part A of the figure shows lab-created plasmids (called expression vectors), made from inserting genes or other genetic elements into a bacterial plasmid. These plasmids were constructed to test whether a gene promoter is functional or not. Recalling the structure of the *Lac* operon, which of the two vectors are negative controls?

A. The top two vectors

B. The bottom two vectors

C. The top and bottom vectors

D. All of them are controls

Answer: C.

The top and bottom vectors. The top vector has no promoter, so luciferase will not be expressed. The bottom vector has a backwards promoter, so luciferase should also not be expressed in bacteria with that plasmid.

31. **Part B of the figure shows light output measured in a luminometer. The GST-28A promoter is known to be functional. According to the graph of light readings, which other promoter is functional?**

A. The Boudicca promoter

B. The Sinbad promoter

C. Both the Boudicca and the Sinbad promoter

D. Neither the Sinbad nor the Boudicca promoter

Answer: B.

The Sinbad promoter. Only the Sinbad promoter has high enough activity that its error bars overlap with the positive control (GST-28A) and do not overlap with the negative control (pGL3-basic).

32. **Part B of the figure shows a surprising, unexpected result. What is this result and what does it indicate?**

A. The Boudicca promoters were nonfunctional. This indicates that the Boudicca retrotransposon does not encode luciferase.

B. The GST-28A promoter had a very large error bar. This indicates that this promoter was not a good choice for a positive control.

C. The Sinbad promoter did not express as much luciferase as the GST-28A promoter. This indicates that the Sinbad promoter is probably nonfunctional.

D. Both the forward and inverted Sinbad promoters were functional. This indicates that this is an unusual class of promoter, known as a bidirectional promoter.

Answer: D.

Both the forward and inverted Sinbad promoters were functional. This indicates that this is an unusual class of promoter, known as a bidirectional promoter. Both the Sinbad forward and Sinbad inverted promoter were equally capable of inducing luciferase expression, and not significantly different from the positive control, since the error bars overlap. This was unexpected, and in fact the inverted promoter was set up as an additional negative control. However, bidirectional promoters, while unusual, do exist.

33. In Part A of the figure, part of the vector plasmid is labelled "amp" for ampicillin resistance—this is where an ampicillin resistance gene has been cloned in. Why has an ampicillin resistance gene been cloned into the vector?

A. So that the researcher can know whether bacteria have been successfully transformed with the plasmid.

B. So that the researcher can make sure the organisms with the plasmid do not become infected with bacteria.

C. So that the plasmids can be used in animals being prophylactically treated with antibiotics.

D. So that the plasmid is protected against ampicillin-resistant bacteria that would interfere with the experiment.

Answer: A.

So that the researcher can know whether bacteria have been successfully transformed with the plasmid. Molecular biologists clone in an antibiotic resistance gene and then grow the transformed bacteria in medium supplemented with the antibiotic. That way, any bacteria that did NOT take up the plasmid is killed, so the remaining bacteria should be transformed bacteria.

Practice Test Two

Questions 34-41. A team of researchers from the USDA and the Centro de Ecologia in Mexico discovered a strain of parasitic wasps that was all-female. These wasps are natural predators of fruit flies, a natural alternative to chemical pesticides. A genetic study indicated that, while the all-female and male and female (bisexual) strain were morphologically similar, the all-female strain may in fact be a different species. After observing that the females of the all-female strain did not mate with the males of the bisexual strain, they did a series of experiments to see whether the behavior of the two strains was statistically different. In the experiments, they used an olfactometer, a multi-armed chamber in which one arm contained bait such as a guava infested with the fruit fly prey of the wasps. They then measured the time spent in the target arm of the device. Questions 34-41 refer to the figure to the right, which describes these experiments. Thelytokous is another name for an all-female strain and arrhenotokous is another name for the bisexual strain.

34. **In this experiment, what is the dependent variable and what is the independent variable?**

   A. The independent variable is the time spent in the target arm of the olfactometer; the dependent variable is the strain of wasp (all-female or bisexual).

   B. The independent variable is the strain of wasp (all-female or bisexual); the dependent variable is the time spent in the target arm of the olfactometer.

   C. The independent variable is which arm of the olfactometer the wasp went to; the dependent variable is the strain of wasp (all-female or bisexual).

   D. The independent variable is the strain of wasp (all-female or bisexual); the dependent variable is which arm of the olfactometer the wasp went to.

Answer: B.

The independent variable is the strain of wasp (all-female or bisexual); the dependent variable is the time spent in the target arm of the olfactometer. The independent variable represents the different conditions being compared; in this case, it is the all-female vs. the bisexual strain. The dependent variable is the variable being measured. In this case, it is time spent in the target arm of the olfactometer.

35. **Part A represents the time it took the wasps to find the guava bait. Infested guavas were tested in one experiment, and non-infested guavas were compared in a separate experiment. Which of the following statements describes the results?**

    A. The all-female wasps found the non-infested guava significantly faster than the bisexual wasps.

    B. The bisexual wasps found the non-infested guava significantly faster than the all-female wasps.

    C. There was no statistical difference in time needed to find the non-infested guava.

    D. Both strains of wasps took longer to find the infested guava than the non-infested guava.

Answer: A.

The all-female wasps found the non-infested guava significantly faster than the bisexual wasps. The all-female wasps took less time, and the difference is statistically significant (as shown by the error bars). Answer D is tempting, but you cannot compare the time taken to find the infested guava with the non-infested guava because these were two separate experiments.

36. **Part B represents the time spent in the arm with the target guava. Infested guavas were tested in one experiment, and non-infested guavas were compared in a separate experiment. Which of the following statements describes the results?**

    A. The arrhenotokous wasps spent more time in the target arm than the thelytokous wasps, for both infested and non-infested guavas.

    B. The thelytokous wasps spent more time in the target arm than the arrhenotokous wasps, for both infested and non-infested guavas.

    C. The arrhenotokous wasps spent significantly more time with the infested guava than the thelytokous wasps, but the thelytokous wasps spend more time with the non-infested guava.

    D. The thelytokous wasps spent significantly more time with the non-infested guava than the arrhenotokous wasps, but the time spent with the infested guava was not statistically different between the two.

Answer: D.

The thelytokous wasps spent significantly more time with the non-infested guava than the arrhenotokous wasps, but the time spent with the infested guava was not statistically different between the two. The time spent with the infested guava was not statistically different, as shown by the overlapping error bars. Thelytokous wasps did spend significantly more time with the non-infested guavas, however.

37. **This type of wasp finds its prey by entering the fruit and swimming through the pulp until it finds a fruit fly larva to parasitize. Part C shows the amount of time spent by the wasps actually swimming in the guava in the target arm. What conclusions can be drawn from the data shown in parts B and C?**

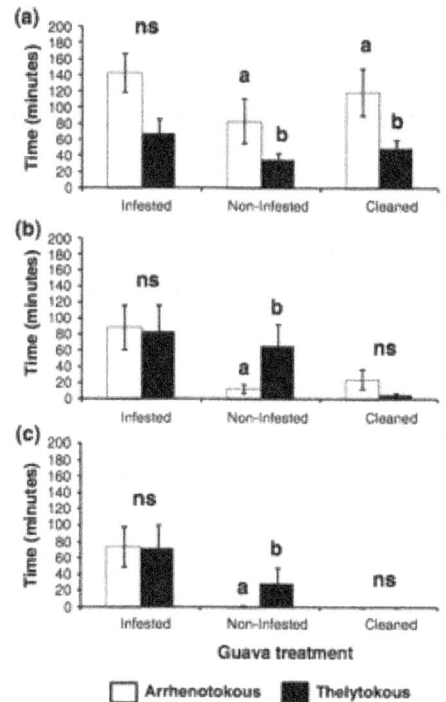

A. The thelytokous wasps behave similarly to the arrhenotokous wasps when prey is present, but spend significantly less time exploring non-infested fruit, which could be a fitness disadvantage.

B. The thelytokous wasps behave similarly to the arrhenotokous wasps when prey is present, but waste more time on non-infested fruit, which could be a fitness disadvantage.

C. The thelytokous wasps prefer non-infested guavas, while the arrhenotokous wasps prefer infested guavas, which could be a fitness advantage for the arrhenotokous wasps.

D. The thelytokous wasps prefer non-infested guavas, while the arrhenotokous wasps prefer infested guavas, which could be a fitness disadvantage for the arrhenotokous wasps.

Answer: B.

The thelytokous wasps behave similarly to the arrhenotokous wasps when prey is present, but waste more time on non-infested fruit, which could be a fitness disadvantage. The two groups are not significantly different in time spent in/around infested guavas—they both spent time in fruit with prey. However, whereas the arrhenotokous wasps spent very little time in or around the non-infested guavas, the thelytokous wasps spent significantly more time there. This could be an evolutionary fitness disadvantage, since they are wasting their time and energy on fruit without prey.

38. **Which of the following represents a reasonable speculation regarding these results— something that cannot be concluded but would be a reasonable hypothesis for further research?**

A. Perhaps the thelytokous wasps are good at smelling guava, but are not very good at smelling fruit fly larvae (thelytokous wasps orient to guava).

B. Perhaps the arrhenotokous wasps are good at smelling fruit fly larvae, but not very good at smelling guava (arrhenotokous wasps orient to fruit fly larvae).

C. Perhaps the arrhenotokous wasps use smell to detect prey, whereas thelytokous wasps use sight (arrenhotokous wasps orient by smell; thelytokous wasps orient by vision).

D. Both A and B.

Answer: D.

Both A and B. Both of these could explain the data shown in the figure.

39. **What experiments could test the hypothesis that the arrhenotokous wasp strain orients to fruit fly larvae whereas the thelytokous wasp strain orients to fruit?**

A. (1) Use the same olfactometer set-up, but with larvae only instead of guava; (2) Use an olfactometer with a non-infested guava in one arm and larvae in another arm.

B. (1) Use the same olfactometer set-up, but with a different fruit infested with the fruit fly larvae; (2) Use an olfactometer with this different fruit in one arm and an infested guava in another arm

C. (1) Use the same olfactometer set-up, but with guavas infested with dead larvae; (2) Use an olfactometer with a guava infested with live larvae in one arm and dead larvae in another arm

D. (1) Use the same olfactometer set-up, but with guava leaves instead of guava fruit; (2) Use an olfactometer with an infested guava in one arm and a mammal in another arm

Answer: A.

(1) Use the same olfactometer set-up, but with larvae only instead of guavas; (2) Use an olfactometer with a non-infested guava in one arm and a larva in another arm. In experiment 1, if the hypothesis were true, the arrhenotokous wasp will spend significantly more time in the larvae-containing arm than in the other arms, whereas the thelytokous wasps would spend about the same amount of time in larvae-containing and non-larvae-containing arms. In experiment 2, if the hypothesis were true, the arrhenotokous wasps would spend significantly more time in the arm with the larvae only and the thelytokous wasps would spend significantly more time in the arm with the non-infested guava.

40. What trophic level do these wasps represent in a cloud forest ecological setting?

A. The first trophic level

B. The second trophic level

C. The third trophic level

D. The fourth trophic level

Answer: C.

The third trophic level. These parasitoid wasps occupy the third trophic level. The first trophic level is the producers/autotrophs (in this case, the guava tree). The second trophic level is consumers of the producers, or herbivores (in this case, fruit flies that eat guavas). The third trophic level is consumers of herbivores (in this case, the parasitoid wasps). The fourth trophic level is consumers of these predators (in this case, an example would be birds that eat the parasitoid wasps).

41. These two wasp strains may or may not be the same species. However, knowing that they are parasitoid wasps, what higher taxonomic category are they in and what category includes the evolutionary ancestor of both the wasps and their fruit fly prey?

A. The wasps are in the order Hymenoptera and both the wasps and the fruit flies are in the class Insecta (phylum Arthropoda)

B. The wasps are in the phylum Hymenoptera and both the wasps and the fruit flies are in the order Arthropoda (class Insecta)

C. The wasps are in the order Arthropoda and both the wasps and the fruit flies are in the class Hymenoptera (phylum Insecta)

D. The wasps are in the class Arthropoda and both the wasps and the fruit flies are in the order Insecta (phylum Hymenoptera)

Answer: A.

The wasps are in the order Hymenoptera and both the wasps and the fruit flies are in the class Insecta (phylum Arthropoda). The Hymenoptera are the bees and wasps. The Insecta are the insects, and this is one class within the phylum Arthropoda, which also includes non-insect animals like lobsters, spiders, and millipedes. You could also deduce this by elimination of the other answers, knowing that phylum is the broadest, most inclusive category, followed by class, followed by order.

42. Which of the following is true about the reproduction of fungi?

A. Most fungi consist of diploid hyphae until it is time for sexual reproduction. At this point, hyphae from two fungal organisms fuse into a dikaryotic organism (two nuclei per cell), which forms sexual sporangia in which the nuclei fuse into one diploid nucleus, which then undergoes meiosis to form diploid spores.

B. Most fungi consist of haploid hyphae until it is time for sexual reproduction. At this point, hyphae from two fungal organisms fuse into a dikaryotic organism (two nuclei per cell), which forms sexual sporangia in which the nuclei fuse into one diploid nucleus, which then undergoes meiosis to form haploid spores.

C. Most fungi consist of haploid hyphae until it is time for sexual reproduction. At this point, hyphae from two fungal organisms fuse into a diploid organism (two nuclei per cell), which forms sexual sporangia which then undergo meiosis to form diploid spores.

D. Most fungi consist of diploid hyphae until it is time for sexual reproduction. At this point, hyphae from two fungal organisms fuse into a haploid organism, which forms sexual sporangia in which the nuclei fuse into one haploid nucleus, which then undergoes meiosis to form diploid spores.

Answer: B.

Most fungi consist of haploid hyphae until it is time for sexual reproduction. At this point, hyphae from two fungal organisms fuse into a dikaryotic organism (two nuclei per cell), which forms sexual sporangia in which the nuclei fuse into one diploid nucleus, which then undergoes meiosis to form haploid spores. You could answer this by knowing about fungi, but you could also answer by elimination on the basis of a general understanding of haploidy, diploidy, and meiosis: A and C are wrong (among other reasons) because if a diploid nucleus undergoes meiosis, the resulting cells will be haploid, not diploid. D is wrong because a haploid nucleus will not undergo meiosis—they do not have two sets of chromosomes to separate into two daughter cells.

43. Which of the following statements about the ecology of fungi is true?

A. Fungi are autotrophs; most are saprophytes and some are photosynthetic.

B. Fungi are heterotrophs; most are herbivores and some are predators.

C. Fungi are heterotrophs; most are parasites and some are predators.

D. Fungi are heterotrophs; most are saprophytes and some are parasites.

Answer: D.

Fungi are heterotrophs; most are saprophytes and some are parasites. Fungi are heterotrophs—they cannot produce their own food through photosynthesis or chemosynthesis. Most are saprophytes (eaters of dead organisms) but some, such as ringworm and athlete's foot, are parasites (living on or in living organisms).

44. In the traditional view of the ability of plants to colonize land, factors such as tissue differentiation have been considered. Another factor that many biologists now consider essential for the transition of plants to land was association of plants with fungi. What is the general name for this type of biological partnership, and what is the specific name for the structures that form from the relationship between soil-dwelling fungi and plants?

A. Parasitism; mycorrhizae

B. Symbiosis; fungorrhizomes

C. Symbiosis; mycorrhizae

D. Parasitism; fungorrhizomes

Answer: C.

Symbiosis; mycorrhizae. Symbiosis is when organisms live together, including in a mutually beneficial relationship (mutualism); parasitism is when one organism benefits but the other is harmed; commensalism is when one benefits and the other gets no benefit and no harm. Therefore, even if you did not know about mycorrhizae, you could nevertheless improve your chances by eliminating A and D.

45. **Which of the following describes the ecological relationship between protozoans and algae?**

A. Algae are consumers and protozoans are producers; together, they provide food for other consumers in the food web.

B. Algae are producers and protozoans are consumers; together, they provide food for other consumers in the food web.

C. Algae are consumers and protozoans are producers; together, they provide food for other producers in the food web.

D. Algae are producers and protozoans are consumers; together, they provide food for other producers in the food web.

Answer: B.

Algae are producers and protozoans are consumers; together, they provide food for other consumers in the food web. Algae, as photosynthetic organisms, are producers. Protozoans are heterotrophic; they cannot produce their own food, and their relationship between protozoans and algae will be that protozoans are consumers of algae. In turn, larger organisms consume the protozoans, and some consume the algae as well.

46. **Diego and Darius are eating lunch. Diego is eating sushi, which contains seaweed, rice, and salmon. Darius is eating a grilled blue cheese and apple sandwich on whole wheat bread. At what trophic levels are Diego and Darius?**

A. Diego is at trophic levels 2 and 4; Darius is at trophic levels 2, 3, and 4.

B. Diego is at trophic levels 3 and 4; Darius is at trophic levels 2 and 3.

C. Diego is at trophic levels 1 and 3; Darius is at trophic levels 2 and 3.

D. Diego is at tropic levels 2 and 3; Darius is at trophic levels 1 and 2.

Answer: A.

Diego is at trophic levels 2 and 4; Darius is at trophic levels 2, 3, and 4. Autotrophs are level 1, herbivores level 2 and so forth. Diego is eating seaweed and rice, so he is level 2. He is also eating salmon, which is a level 3 organism (predator of smaller animals), putting him at trophic level 4. Darius is also at level 2 because he is eating wheat and apple, and at level 3 because he is eating milk in the form of cheese; milk counts as an herbivore (cow). In addition, Darius is eating the *Penicillium* fungus that consumed some of the milk to make the cheese. The fungus is at level 3, since milk counts as level 2. By eating the *Penicillium* in the blue cheese, Darius is therefore at level 4 as well. *You could quickly eliminate answers C and D because they include trophic level 1, autotrophs.*

47. **Which of the following pairs of organisms can form a relationship that allows the earliest colonization of a barren environment, such as a new volcanic island?**

A. An insect and a moss

B. A fungus and an algae

C. A moss and a fungus

D. An insect and a protozoan

Answer: B.

A fungus and an algae. Lichens consist of fungus and algae living symbiotically. In this relationship, lichens can colonize very barren environments, since the algal partner can photosynthesize, producing food energy, and the fungus can absorb water and minerals from the environment, even if it is sheer rock.

Questions 48-51 refer to the following cladogram:

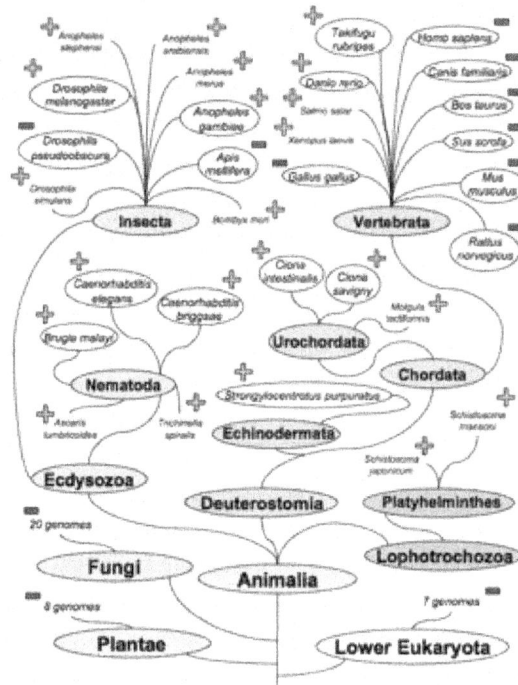

48. According to this "tree of life" view of evolution, animals are divided into three major groups, or superphyla. Which of these three groups includes humans?

A. Vertebrata

B. Chordata

C. Lophotrochozoa

D. Deuterostomia

Answer: D.

Deuterostomia. The three major groups of animals in this classification are the Ecdysozoa, the Deuterostomia, and the Lophotrochozoa. Humans are within the Deuterostomia. A and B do not represent superphyla; they are at much less basal branching levels.

49. **The term "Deuterostomia" means "mouth second." Why is this name used for this group?**

    A. During the embryonic development of the Deuterostomia, the second gastrulation of the hollow zygote takes place at the mouth.

    B. During the embryonic development of the Deuterostomia, gastrulation of the hollow blastula begins at the anus and ends at the mouth.

    C. During the embryonic development of the Deuterostomia, gastrulation of the hollow blastula begins at the feet and ends at the mouth.

    D. During the embryonic development of the Deuterostomia, gastrulation of the hollow gastrula begins at the anus and ends at the mouth.

    Answer: B.

    During the embryonic development of the Deuterostomia, gastrulation of the hollow blastula begins at the anus and ends at the mouth. The blastula is the hollow ball of cells, an early stage of embryonic development, that begins to invaginate and eventually forms a sort of tunnel through the ball of cells, called the gastrula. The process, called gastrulation, begins at the mouth and ends at the anus in protosomes (= "mouth first"); and begins at the anus and ends at the mouth in deuterostomes.

50. **After this cladogram was published, one of the three superphyla was divided into two separate superphyla. The new superphylum is the Platyzoa, and includes flatworms. Of the animals in the cladogram above, which genus would be moved to this new superphylum?**

    A. *Caenorrhabditis*

    B. *Trichinella*

    C. *Schistosoma*

    D. *Ascaris*

    Answer: C.

    Schistosoma. *Schistosoma* is the only flatworm genus in the cladogram. *Caenorrhabditis*, *Trichinella*, and *Ascaris* are all roundworms. Even if you do not know the genera, it is clear from the figure that all three wrong answer genera are from the same phylum, the Nematoda, so if you know that at least one of them is a roundworm, it will be clear that all three are wrong answers.

51. Which of the following groups contains spiders?

A. The Ecdysozoa

B. The Lophotrochoza

C. The Insecta

D. The Echinodermata

Answer: A.

The Ecdysozoa. Spiders, class Arachnida, are from the phylum Arthropoda, in the Ecdysozoa. A general understanding of animal evolution, together with a look at the branching illustrated in the figure, should allow the elimination of the other three answers.

52. A certain protein is destined to be transported to another part of the body. Which of the following will be involved in packaging the protein after gene expression?

A. The Golgi apparatus

B. The rough endoplasmic reticulum

C. The ribosome

D. Protein packagase, a transmembrane protein

Answer: A.

The Golgi apparatus. While ribosomes on the rough endoplasmic reticulum are involved in creating a protein, the Golgi apparatus is the organelle involved in packaging the protein for transport.

53. Cytochrome c oxidase is a large, transmembrane protein that is an integral part of the electron transport chain. Which of the following describes cytochrome c oxidase?

A. Hydrophobic polypeptide with nonpolar amino acids to interact with water, with metal atoms

B. Hydrophilic polypeptide with polar amino acids to interact with water, with cytochrome atoms

C. Alternating hydrophobic and hydrophilic polypeptide regions, with helium atoms

D. Alternating hydrophobic and hydrophilic polypeptide regions, with metal atoms

Answer: D.

Alternating hydrophobic and hydrophilic polypeptide regions, with metal atoms. Transmembrane proteins contain both hydrophobic and hydrophilic regions, allowing the protein to be embedded in a cell membrane. (The hydrophobic regions are embedded inside the lipid bilayer, while the hydrophilic regions stick out on one or the other side of the membrane.) In addition, as a component of the electron transport chain, cytochrome c must be able to engage in oxidation and reduction reactions, which will require metal atoms.

54. **Noncoding RNA is RNA that will not be used as a template for building a polypeptide. Which of the following are examples of noncoding RNA?**

A. ribosomal RNA, messenger RNA, and small nuclear RNA

B. messenger RNA, tRNA, and ribosomal RNA

C. micro RNA, tRNA, ribozymes

D. small nuclear RNA, cDNA, and ribozymes

Answer: C.

micro RNA, tRNA, ribozymal RNA. All three of these types of RNA are noncoding RNA. A and B are wrong because they include messenger RNA, which is used as a template for building a polypeptide. D is wrong because it includes cDNA, which is not a type of RNA at all.

55. **Euglena are single-celled organisms that show both plant- and animal-like properties. They can photosynthesize, but also eat other organisms and can move through their freshwater environment using whip-like cellular appendages. Knowing this, which of the following cellular structures do Euglena have?**

A. Chloroplasts and a cell wall

B. Flagella and chloroplasts

C. Flagella and gap junctions

D. Chloroplasts and plasmodesmata

Answer: B.

Flagella and chloroplasts. Euglena move using flagella and photosynthesize using chloroplasts. You could also eliminate A because a cell wall would prevent movement, and C and D because Euglena are single-celled, so they will not use cell-cell communication structures like gap junctions and plasmodesmata.

56. **Some neurons are very long, for example, neurons that extend from the spine to muscles in your feet. To enable a signal to travel this length quickly enough to allow efficient movement, the axons are coated with insulating cells that allow the signal moving down the axon to move in a jumping/sliding motion called saltatory conduction. What are these insulating cells called, what substance surrounds the axon, and what happens in the gap in between them?**

A. Adipocytes, myelin, membrane potential

B. Neurons, glia, action potential

C. Neuroglia, myelin, action potential

D. Microglia, myelin, membrane potential

Answer: C.

Neuroglia, myelin, action potential. Glial cells, or neuroglia, include the Schwann cells that make the myelin sheaths that surround peripheral nervous system axons to allow saltatory conduction. In between each myelin-coated segment, at gaps called the nodes of Ranvier, a new action potential is generated, generating an electrical signal that is quickly conducted down the myelinated segment to the next node of Ranvier.

57. In order to colonize land, plants had to develop separate systems for obtaining sunlight and obtaining water and nutrients. Seaweeds, in contrast, live in water, which allows the whole plant surface to have access to both sunlight and water/minerals. Knowing this, what is the likely evolutionary history of sea grass, flowering plants that live completely underwater in shallow marine environments like coral reefs?

A. Nonvascular land plants evolved from algae. Then, some of these eventually evolved into flowering land plants. Some of these then evolved to be able to survive underwater in shallow marine habitats.

B. Nonvascular marine plants evolved from algae. Then, some of these eventually evolved into vascular marine plants. Some of these then evolved into flowering marine plants.

C. Multicellular algae (seaweed) evolved from unicellular algae. Then, some multicellular seaweeds evolved into flowering seaweeds. Then, some of these gained a grassy morphology optimized for shallow marine habitats.

D. Multicellular algae (seaweed) evolved from unicellular algae. Next, some of these evolved into nonvascular marine plants. Then, some of these evolved into flowering plants with a grassy morphology optimized for shallow marine habitats.

Answer: A.

Nonvascular land plants evolved from algae. Then, some of these eventually evolved into flowering land plants. Some of these then evolved to be able to survive underwater in shallow marine habitats. Since the seagrass is a flowering plant, it must be a descendent of land flowering plants, which evolved reproductive structures optimized for reproduction in a land environment. Going backwards in evolutionary time, the flowering land plants evolved from simpler plants which evolved from algae.

58. **During the Calvin cycle of photosynthesis, plants create glucose using the energy in sunlight to build glucose molecules. What is the net reaction of this process?**

A. $6 CO_2 + 6 O_2 =>$ light energy $=> C_6H_{12}O_6 + 6 H_2O$

B. $6 CO_2 + 6 H_2O =>$ light energy $=> C_6H_{12}O_6 + 6 O_2$

C. $C_6H_{12}O_6 + 6 O_2 =>$ light energy $=> 6 CO_2 + 6 H_2O$

D. $C_6H_{12}O_6 + 6 O_2 => 6 CO_2 + 6 H_2O +$ energy

Answer: B.

$6 CO_2 + 6 H_2O =>$ light energy $=> C_6H_{12}O_6 + 6 O_2$. The plant uses light energy to build glucose. Only A and B include glucose as a product. A is a nonsensical chemical reaction.

59. **Hookworms are roundworms that hook into the intestinal lining and suck blood for their nourishment. What type of organism are hookworms?**

A. Protozoan parasite

B. Protozoan symbiont

C. Metazoan parasite

D. Metazoan mutualist

Answer: C.

Metazoan parasite. Hookworms are metazoan parasites. You would know that they are not protozoan, since the question says they are roundworms (eliminate A and B). You would know that they are parasites, not mutualists (each benefitting the other), so that eliminates answer D.

60. **The protozoan parasitic amoeba *Entamoeba histolytica* causes a serious disease, but molecular studies have found that what was thought to be a single organism is in fact two: the pathogenic *Entamoeba histolytica* and a non-pathogenic commensal, *Entamoeba dispar*. Which animal is analogous in lifestyle to *Entamoeba dispar*?**

A. *Plasmodium falciparum*, the protozoan that causes malaria

B. *Demodex folliculorum*, a mite that lives on the skin of almost all humans and eats oil from sebaceous glands

C. Intestinal bacteria, which play a role in synthesizing vitamin B and vitamin K as well as metabolizing bile acids, sterols and xenobiotics

D. *Paragonimus westermani*, a lung fluke that causes the human disease paragonimiasis

Answer: B.

Demodex folliculorum, a mite that lives on the skin of almost all humans and eats oil from sebaceous glands. Of these, only Demodex is a commensal like Entamoeba dispar. Plasmodium and Paragonimus are parasites, and the intestinal bacteria described are mutualists.

61. Which of the following describe communication between two neurons?

A. A neurotransmitter is released into the synapse, it binds a receptor on the postsynaptic neuron, and this generates an action potential.

B. A neurotransmitter is released into the synapse, it binds to a receptor on the postsynaptic neuron, and this generates a membrane potential.

C. A hormone is released into the synapse, it binds to a receptor on the postsynaptic neuron, and generates an action potential.

D. A cytokine is released into a gap junction, where it travels to the postsynaptic neuron, and generates a membrane potential.

Answer: B.

A neurotransmitter is released into the synapse, it binds to a receptor on the postsynaptic neuron, and this generates a membrane potential. The molecule that is used to signal between two neurons is a neurotransmitter. It will generate a membrane potential, and if that membrane potential is large enough, it will generate an action potential.

62. Male mosquitoes have large, feathery antennae and mouthparts adapted to a diet of nectar, whereas female mosquitoes have much smaller antennae and long, syringe-like mouthparts. Why?

A. Males can survive on flower nectar, but females require the extra protein of a blood meal in order to lay eggs. Males with bigger antennae are better at detecting females.

B. Females can survive on flower nectar, but they require the extra protein of a blood meal in order to lay eggs. Females with bigger antennae are better at attracting males.

C. There is no adaptive advantage to either trait. The evolution of these traits is the result of genetic drift.

D. The long, sucking mouthparts of the females enable them to obtain more calories than afforded by flower nectar alone. The large antennae of males is the result of higher levels of testosterone.

Answer: A.

Males can survive on flower nectar, but females require the extra protein of a blood meal in order to lay eggs. Males with bigger antennae are better at detecting females. Dramatically different traits between the sexes generally are functional (making C highly unlikely). The function of the female mouthparts is to provide extra protein for eggs; the function of the large male antennae is to find females.

63. **Which molecule allows mosquitoes to detect humans or other mammals that can provide a blood meal?**

A. O_2

B. ATP

C. CO_2

D. $C_6H_{12}O_6$

Answer: C.

CO_2. Humans and other mammals are constantly undergoing aerobic respiration, and therefore constantly emitting the waste product of aerobic respiration, which is CO_2. This is the only molecule of the four answers that is emitted by humans (a biochemical product) rather than used as a reactant (O_2 and $C_6H_{12}O_6$), or created but then stored and used (ATP).

Part B. Grid-in Questions Answer Key

1. A research lab is using mutant mice for its experiments to determine the effects of a certain gene deficiency on blood pressure. Each month, 200 mice are born and 45 mice die. The mouse colony currently has 630 mice. What is the growth rate of the colony each month? Round your answer to the hundredth.

   Answer: 0.25

   Explanation: Rate= (births-deaths)/N(population). Therefore: 200-45= 155; then 155/630=0.246

2. A green-fluorescent protein exists that will localize to the mitochondrial membrane. A researcher would like to visualize the mitochondria in murine macrophages. She has a stock solution of the stain at a concentration of 2µM and the protocol calls for a final concentration of 50nM in a volume of 200 microliters. How many microliters of stock solution does the researcher need?

   Answer: 5

   Explanation: 50nM (what we require)/2000nM (stock) * 200µl (volume we require) =5µl

3. Evaporation of water from leaves of plants to the environment is called transpiration. The graph below shows the transpiration rates of an avocado tree throughout the year. How many liters of water does an avocado tree lose during the summer (June through August)?

   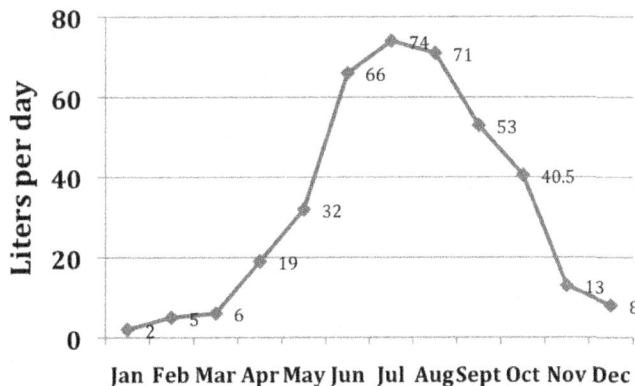

   Answer: 6475

Explanation: 30 days in June * 66 = 1980
31 days in July * 74 = 2294
31 days in August *71 = 2201
(1980+2294+2201)=6475

4. **In a particular species of sloth, there is a recessive allele of a gene that results in an unusually long tail. This allele occurs at a frequency of 0.527 in the population. According to the Hardy-Weinberg equation, what is the frequency of long-tailed sloths? Round your number to the nearest thousandth.**

Answer: 0.278

Explanation: One of the Hardy-Weinberg Equations is $p^2 + 2pq + q^2$, where q is the frequency of the homozygous recessive gene. Therefore, $0.527^2 = 0.2777$

5. **During vasculogenesis (the development of blood vessels), cells called angioblasts are differentiated into endothelial cells. These endothelial cells migrate in response to growth factors to their correct location and will become the inner lining of capillaries. If a single angioblast differentiates into an endothelial cell, and the subsequent rate of division is once every 4 minutes, how long will it take to produce enough cells to line a capillary requiring 262, 144 cells? Report you answer in hours.**

Answer: 1.2 hours

Explanation: In general, you would use the equation
$$2^n = 262,144$$

where 2 represents the number of cells produced after each division, and n is the number of divisions occurring – in this case. Therefore, the number of cell divisions would be:

$$n = \text{Log}_2 262,144$$

Solving this with the help of a calculator, we will get:

$$n = 18$$

Since we have a cell division every 4 minutes, we need to multiply 18 by 4 to get the number of minutes, 72. To get the number of hours, divide 72 by 60, to get 1.2 h

6. Erythropoietin is a hormone that controls red blood cell production.

Translation of the erythropoietin gene results in a protein containing 193 amino acids. This amino acid chain is later cleaved to produce a chain that is 166 amino acids. How many ATP molecules are required for the translation of erythropoietin?

Answer: 772

Explanation: It takes the energy from about 4 ATP molecules in the translation of each amino acid. If erythropoietin has 193 amino acids encoded in its gene, then (193 * 4) = 772 ATP molecules are required.

1. Imagine that you are digesting a grilled cheese sandwich. Name and briefly describe the function of three enzymes that will be used to break down the sandwich, and name and briefly describe the function of three organs and their physiological function involved in the digestive process. (Consider only digestion and not absorption of nutrients or excretion of waste.)

Model Answer:

You could choose any three enzymes and any three organs from the following (or any other, with a well-argued defense). On the free response section of the AP exam, you will not lose points for extra information, so on a question like this, if you have time, list more than the required number of items.

Enzymes:

- **Salivary amylase:** breaks down carbohydrates
- **Lingual lipase:** breaks down fats
- **Pepsinogen:** breaks down proteins
- **Trypsin:** breaks down proteins
- **Pancreatic lipase:** breaks down fats
- **Pancreatic amylase:** breaks down carbohydrates
- **Lysozyme:** kills microbes in the mouth
- **Haptocorrin:** binds to vitamin B12 to protect it from the acidic conditions of the stomach (not used to breakdown any specific nutrient, but it is involved in the proper breakdown of the food as a whole)
- **Gastric Lipase:** breaks down fats
- **Chymotrypsinogen:** breaks down proteins
- **Carboxypeptidase:** breaks down proteins
- **Phospholipase:** breaks down phospholipids
- **DNAse:** breaks down DNA
- **RNAse:** breaks down RNA
- **Maltase:** converts maltose into glucose
- **Lactase:** converts lactose into glucose and galactose (many Asians, Middle-Easterners, and older people are deficient in this enzyme, leading to bloating and pain when they drink milk or milk products)
- **Sucrase:** converts sucrose (table sugar) into glucose and fructose

There are many more! Try to list more than the question asks for if you have time. Also, since you will not get any points off for wrong answers, and can gain credit for partially correct answers, if you do not know a specific answer, write what you do know. For example, you could write:

- proteases and peptidases split proteins into small peptides and amino acids.
- lipases split fat into three fatty acids and a glycerol molecule.
- amylases split long carbohydrates such as starch into simple sugars such as glucose.
- nucleases split nucleic acids into nucleotides.

Organs:
- **Mouth/oral cavity:** breaks down large chunks of food by chewing, begin breaking down carbohydrates and fats through enzymes in saliva, kills microorganisms, protective/ preparative functions such as binding vitamin B12 with haptocorrin to protect it from stomach acids
- **Stomach:** secretes hydrochloric acid to kill microbes and pepsinogen to begin to break down proteins; converts food into chyme
- **Pancreas:** secretes enzymes into the small intestine to continue the digestion of carbohydrates, fats, and proteins
- **Gallbladder:** secretes bile into the small intestine to aid digestion and absorption of fats
- **Small intestine:** digestive enzymes continue to break down carbohydrates, fats, and proteins

2. **Describe the role of oxygen in cellular respiration. What will occur if oxygen is not available?**

Model Answer:

Oxygen plays an important role in cellular respiration because the electrons are donated to oxygen at the end of the electron transport chain. In the process, a gradient of H^+ ions forms across the mitochondrial membrane, and the potential energy of these H^+ ions is harvested by ATP synthase, in theory generating 36 molecules of ATP for the whole process. In the absence of oxygen, the electron transport chain will not proceed, and cells can only survive by fermentation. Fermentation does not generate enough energy to meet the needs of large, multicellular organisms. Oxygen plays a crucial role in aerobic respiration because it drives the processes to completion.

The net reaction of aerobic respiration is essentially the combustion of organic molecules,

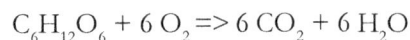

$$C_6H_{12}O_6 + 6\,O_2 => 6\,CO_2 + 6\,H_2O$$

except that instead of the instant release of energy seen in a fire, the energy release is slowed by the electron transport chain, which allows it to be harvested by the cell.

3. **Explain the differences between microtubules and microfilaments.**

Model Answer:

Microtubules and microfilaments are both protein polymer structures that are found in the cytoskeleton. Microtubules are the largest fibers and are made of tubulin. Cilia and flagella are made of microtubules. Sperm, fallopian tubes, centrioles, and the trachea all contain these and use them for locomotion. Microtubules also form mitotic spindles during cell division. Microfilaments are smaller than microtubules, made of actin and myosin, and facilitate cellular movement like cell migration, cytoplasmic streaming and endocytosis.

4. **A deficiency of proteins responsible for nuclear membrane reassembly prevents cells from properly undergoing mitosis. List the stages of mitosis and explain what will happen in the case of such a protein deficiency.**

Model Answer:

Mitosis can be divided into the following stages: prophase, metaphase, anaphase, and telophase. It begins with prophase, when the chromatin condenses to become visible chromosomes. Next, the nucleolus disappears and the nuclear membrane breaks apart. Then, mitotic spindles made of microtubules are formed to eventually pull the chromosomes apart. Also during prophase, the cytoskeleton breaks down and the centrioles push the spindles to the poles at opposite ends of the cell. In addition, the nuclear membrane fragments, to allow the spindle microtubules to interact with the chromosomes, while the kinetochore fibers attach to the chromosomes at the centromere region. Metaphase begins when the centrosomes are at opposite ends of the cell and the centromeres of all the chromosomes are aligned with one another. During anaphase, the centromeres split in half, and the homologous chromosomes separate. The chromosomes are pulled to the poles of the cell, with identical sets at either end. The last stage of mitosis is telophase when two nuclei form, each with a full set of DNA that is identical to the parent cell. The nucleoli become visible and the nuclear membrane reassembles. If the nuclear membrane cannot reassemble, the DNA will have been copied properly, but the copied sets of DNA cannot be packaged into nuclei.

5. **Explain the differences between meiosis and mitosis.**

Model Answer:

Mitosis occurs in somatic cells and produces identical diploid cells through copying both sets of chromosomes. (In humans, each somatic daughter cell will have 46 chromosomes.) This takes place as part of a single cell division event, and results in two diploid daughter cells.

Meiosis occurs in germ cells, or sex cells, to produce gametes, which are haploid cells. The two sets of chromosomes are copied, but then divided again, so that the resulting cells have only one set of chromosomes. (In humans, each gamete daughter cell will have 23 chromosomes.) The two cell division events associated with meiosis results in four haploid daughter cells.

6. **Describe the physiological aspects of glucose processing during cellular respiration, with an emphasis on the locations in which the process occurs.**

Model Answer:

Glucose begins to get processed during glycolysis in the cytoplasm. There, it is converted to two molecules of pyruvate with a net yield of 2 ATP and 2 NADH per glucose molecule.

Through active transport, pyruvate moves into the mitochondrion and is converted into acetyl CoA. In the inner mitochondrial matrix, in the presence of oxygen, it is processed through the Krebs cycle, resulting in the production of 2 more ATP molecules per original starting amount of glucose molecules. Processing through a series of enzymes called cytochromes in the inner mitochondrial membrane, known as the electron transport chain, results in a gradient of H^+ ions in the intermembrane space (between the inner mitochondrial membrane and the outer mitochondrial membrane). The flow of these H^+ ions back across the inner mitochondrial membrane into the mitochondrial matrix, through the transmembrane protein ATP synthase, results in the synthesis of (theoretically) 36 ATP molecules per glucose molecule.

7. **Photosynthesis is crucial to the survival of plants, algae, and cyanobacteria.**

   (A) Outline the light-dependent reactions of photosynthesis.

   (B) Explain how temperature could affect these processes.

   (C) Graph the expected results.

Model Answer:

A:
- Pigment in Photosystem II absorbs light.
- This absorption of light produces an excited electron.
- The electron is passed along a series of carriers to higher levels.
- These electrons arrive to their final electron acceptor $NADP^+$, and the reduction of $NADP^+$ generates NADPH.
- This reaction in Photosystem II provides electrons for Photosystem I.
- Photolysis of water produces free hydrogen ions and oxygen as a by-product in a process called non-cyclic photophosphorylation.
- In cyclic photophosphorylation, the electron returns to chlorophyll.
- The generation of ATP occurs when H^+ is pumped across the thylakoid membrane through the ATP synthase protein.

B:

The rate of photosynthesis will likely increase as the temperature increases as a result of the higher rate of enzyme catalysis of light-dependent or light-independent reactions in the Calvin cycle. At very high temperatures, however, the enzymes required for photosynthesis, including rubisco, could denature, negatively affecting the rates of photosynthesis.

C:

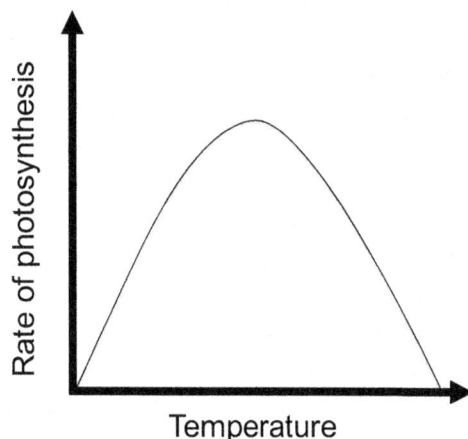

8. **In order to understand the processes of cellular respiration in muscle cells grown in a laboratory cell culture, the cells were lysed and fractionated to isolate the mitochondria from the cytoplasm. The fractionated samples were treated with glucose or pyruvate for 1 hour, and at the end of the incubation period, each sample was analyzed for the presence of lactic acid or carbon dioxide. The questions below should be answered based on the results presented here:**

Cell fraction	CO2	Lactic Acid
Mitochondria incubated with glucose	Absent	Absent
Mitochondria incubated with pyruvate	Present	Absent
Cytoplasm incubated with glucose	Absent	Present
Cytoplasm incubated with pyruvate	Absent	Present

a. What does the presence of lactic acid in the samples suggest about which process is occurring in each particular cell fraction?

b. Why was carbon dioxide not produced by the cytoplasmic fraction incubated with glucose?

c. Why did the cytoplasmic fraction produce lactic acid in the presence of both glucose and pyruvate?

d. Why did the mitochondrial fraction produce carbon dioxide in the presence of pyruvate but not in the presence of glucose?

e. Draw the chemical reaction occuring in each scenario.

Model Answer:

a. The presence of lactic acid suggests anaerobic conditions. Pyruvate, the end product of glycolysis, undergoes fermentation and produces lactic acid.

b. Since glycolysis occurs in the cytoplasm, glucose can be converted there, but only to pyruvate which cannot continue to the Krebs cycle, so carbon dioxide is not an end product of glycolysis.

c. Glycolysis occurs in the cytoplasm, with pyruvate formed as the end product. If there is no oxygen to consume, the cellular respiration process will cease, resulting in anaerobic respiration. Fermentation is a type of anaerobic respiration that converts pyruvate to lactic acid. Therefore, cytoplasmic fractions treated with glucose or pyruvate will only result in the formation of fermentation products.

d. Glucose is broken down to pryuvate by glycolysis, which occurs in the cytoplasm. Therefore, no breakdown products could be observed by the addition of glucose to mitochondria. Pyruvate is a product of glycolysis that is transported to the mitochondria via active transport where it undergoes processing in the Krebs cycle, which produces carbon dioxide.

e. 1. Mitochondria incubated with glucose: no reaction

   2. Mitochondria incubated with pyruvate:
      $$2 \text{ Pyruvate} + 2NADH \rightarrow 2 \text{ Acetyl CoA} + CO_2 + 2NAD^+$$

   3. Cytoplasm incubated with glucose:
      $$\text{Glucose} + 2ATP + 2NAD^+ \rightarrow 2 \text{ Pyruvate} + 4ATP + 2NADH$$

   4. Cytoplasm incubated with pyruvate
      $$2 \text{ Pyruvate} + 2NADH \rightarrow 2 \text{ lactic acid} + 2NAD^+$$

AP

The Advanced Placement® program is designed to offer students college credit while still in high school. The more than 30 AP courses culminate in an intensive final exam given every year in May.

Successful completion of a course and a passing score on the exam not only provides students with a deep sense of accomplishment, but also gives them a jumpstart on their college careers. AP credit is almost universally accepted by post-secondary schools, however each school has different guidelines as to what scores they will accept.

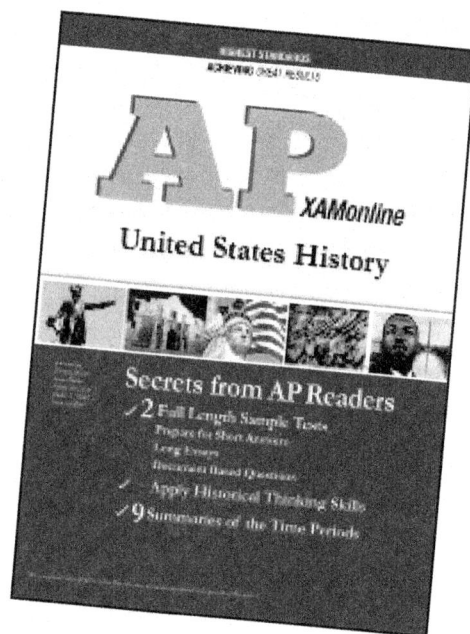

AP US History
ISBN 978-1-60787-552-9 $21.99

AP US Government and Politics
ISBN 978-1-60787-601-4 $21.99

AP Biology
ISBN 978-1-60787-553-6 $21.99

AP Calculus
ISBN 978-1-60787-555-0 $21.99

AP Chemistry
ISBN 978-1-60787-554-3 $21.99

AP Psychology
ISBN 978-1-60787-556-7 $21.99

AP English
ISBN 978-1-60787-557-4 $21.99

AP Spanish
ISBN 978-1-60787-558-1 $21.99

AP Macroeconomics/Microeconomics
ISBN 978-1-60787-585-7 $21.99

TO ORDER

XAMonline.com

or **amazon** or **BARNES & NOBLE** BOOKSELLERS

CPSIA information can be obtained
at www.ICGtesting.com
Printed in the USA
BVOW04s1939110817
491837BV00027B/361/P